Prais
Dispatches from

"This must-read and eminently readable study takes readers on a voyage through the seemingly intractable problems in the South China Sea to reveal how through environmental collaboration competing nations can adopt trust and science-driven peace building measures that can help reduce the risks of conflict."

—Carla Freeman, Senior Expert China,
United States Institute of Peace

"The elegantly written memoir chronicles the gradual erosion of the marine biodiversity of the South China Sea, a victim of the power politics in the region."

—James Kraska, Chair and Charles H. Stockton
Professor of International Maritime Law,
Stockton Center for International Law,
US Naval War College

"James Borton provides a personal and thoughtful new book about Vietnamese fishermen who are caught in the middle of sovereignty disputes and environmental security issues."

—Binh Lai (Ph.D.), Deputy-General,
East Sea (South China Sea) Institute

"A timely book that grapple with some of the biggest issues of our time: finding meaningful solutions to the existential planetary crisis posed by climate change and the intensifying geopolitical competition in the South China Sea. An insightful read for anyone interested in the future of the Indo-Pacific."

—Dr. Manali Kumar Editor-in-Chief 9Dashline

"This is a book about hope and the future of marine biodiversity and sustainability in the South China Sea. It is a major contribution to an understudied field."
—Larry Berman, Professor Emeritus,
University of California, Davis

"Dispatches from the South China Sea is a timely and thought provoking book that explains the intricate relations and exposes the devastating environmental impacts that bring concern to all nations from in the Indo-Pacific region."
—Rear Admiral Scott Sanders,
United States Navy (Retired)

"Borton combines his own expert knowledge of the region with a broad range of perspectives. It is an once an arresting yet hopeful book—one that ought to change how even the most informed readers think about the South China Sea."
—Dr. Peter Harris, Department of Political Science
at Colorado State University, and editor of
"Indo-Pacific Perspectives" section of the
Journal of Indo-Pacific Affairs

"Journalist James Borton's elegant writing echoes Rachel Carson as he blows the whistle on a powerful environmental and human catastrophe of coral reef destruction, overfishing, illegal fishing and murder on the open sea."
—Skye Moody, author of
Washed Up, The Curious Journeys of Flotsam and Jetsam

"Maritime protein (fish, squid, and crabs) is a critical and often overlooked element of competition and contention in the South China Sea. In bringing this to the fore, and drawing on the perspective of the fishermen and marine scientists, Borton adds an important component to our understanding of the South China Sea and management of regional tensions."
—Rodger Baker,
Senior Vice President Strategic Analysis, Stratfor

"In this bold, personal, and unapologetic narrative, Borton uses compelling stories from fishermen and marine scientists to offer key insights on peace-building through science cooperation."
—Severine Autesserre, author of *Peaceland*
and *The Frontlines of Peace*

DISPATCHES FROM
THE SOUTH CHINA SEA

Also by James Borton

Venture Japan
The Art of Medicine in Metaphors
The South China Sea Challenges and Promises
Islands and Rocks in the South China Sea: Post Hague Ruling

DISPATCHES FROM THE SOUTH CHINA SEA

NAVIGATING TO COMMON GROUND

JAMES BORTON

Universal-Publishers
Irvine • Boca Raton

Dispatches from the South China Sea: Navigating to Common Ground

For permission to photocopy or use material electronically from this work, please access www. copyright.com or contact the Copyright Clearance Center, Inc. (CCC) at 978-750-8400. CCC is a not-for-profit organization that provides licenses and registration for a variety of users. For organizations that have been granted a photocopy license by the CCC, a separate system of payments has been arranged.

Universal Publishers, Inc.
Irvine • Boca Raton
USA • 2022
www.Universal-Publishers.com

ISBN: 978-1-62734-370-1 (pbk.)
ISBN: 978-1-62734-371-8 (ebk.)

Typeset by Medlar Publishing Solutions Pvt Ltd, India
Cover design by Wesley Strickland

Library of Congress Cataloging-in-Publication Data

Names: Borton, James W., author.
Title: Dispatches from the South China Sea : navigating to common ground /
   James Borton.
Description: Irvine : Universal Publishers, 2022. | Includes
   bibliographical references and index.
Identifiers: LCCN 2021043890 (print) | LCCN 2021043891 (ebook) |
   ISBN 9781627343701 (paperback) | ISBN 9781627343718 (ebook)
Subjects: LCSH: Fishery resources--South China Sea. | Marine biodiversity
   conservation--South China Sea. | Sustainable fisheries--Government
   policy--Southeast Asia. | Fisheries--South China Sea. |
   Fisheries--Environmental aspects--South China Sea. | Fishery
   policy--Southeast Asia. | Fishery policy--International cooperation.
Classification: LCC SH214.56 . B67 2022 (print) | LCC SH214.56 (ebook) |
   DDC 333.95/60916472--dc23
LC record available at https://lccn.loc.gov/2021043890
LC ebook record available at https://lccn.loc.gov/2021043891

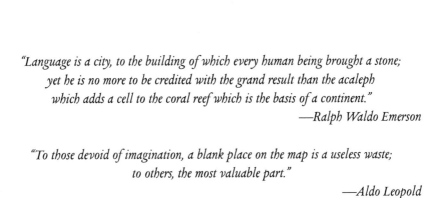

*"Language is a city, to the building of which every human being brought a stone;
yet he is no more to be credited with the grand result than the acaleph
which adds a cell to the coral reef which is the basis of a continent."*
—*Ralph Waldo Emerson*

*"To those devoid of imagination, a blank place on the map is a useless waste;
to others, the most valuable part."*
—*Aldo Leopold*

Map 1. Maritime Claims in the South China Sea

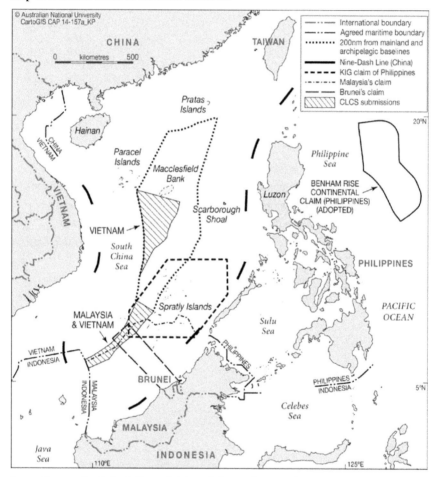

Source: Cartography Unit, College of Asia and the Pacific, Australian National University.
Note: CLCS stands for Commission of the Limits of the Continental Shelf. https://
www.researchgate.net/publication/304630611_China%27s_Ambition_in_the_South_
China_Sea_Is_a_Legitimate_Maritime_Order_Possible/figures?lo=1.

Map 2. Competing Claims in the South China Sea

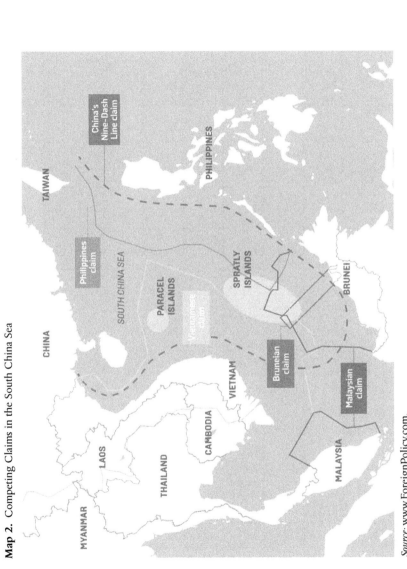

Source: www.ForeignPolicy.com.
Permission granted courtesy of ForeignPolicy.com.

Map 3. South China Sea

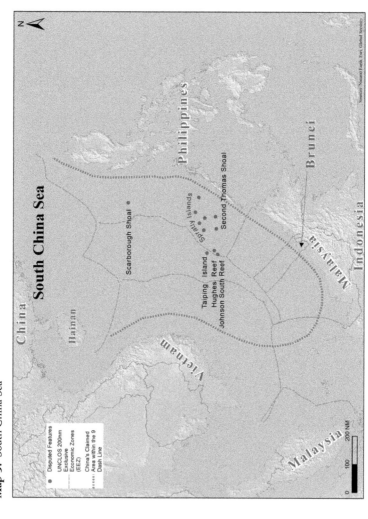

Source: Graphic courtesy Law of the Sea: A Policy Primer project, The Fletcher School of Law and Diplomacy, Tufts University. Copyright 2017 by Trustees of Tufts College, all rights reserved.

Table of Contents

Part II

Ecological Politics

Part III

Science Cooperation and Diplomacy

Appendices

Preface

O n the world atlas, the South China Sea appears as a speck of blue amid the coil of atolls, islands, peninsulas and rocks that comprise Southeast Asia between the Indian and Pacific oceans. Yet, this 1.4 million-square-mile expanse is the major artery for more than $5 trillion shipped annual cargo and serves up approximately 12 percent of the world's fish catch. But it is in this unique natural marine laboratory and gateway to deep-sea ambitions that an environmental crime scene remains unsolved. The South China sea faces a serious problem because of the mounting environmental degradation due to climate change, ocean acidification, plastics pollution, reclamations, overfishing, and population pressures from all neighboring states.

It's my hope that this book will raise awareness for the conservation of marine biodiversity and sustainability of fisheries that can no longer be ignored. The impact of continuous coastal development, reclamation, destruction of corals, overfishing and increased maritime traffic places all of us on the front lines. Marine biologists, who share a common language that cuts across political, economic and social differences, recognize that the structure of a coral reef strewn with the detritus of perpetual conflict represents one of nature's cruel battlefields. The sea's remarkable coral reefs, which provide food, jobs and protection against storms and floods, have suffered unprecedented rates of destruction in recent decades.

At a time when the role of science collaboration is most needed to address the Covid-19 pandemic, there's the potential for the virus to trigger quantifiable conservation strategy in the safeguards taken for biodiversity outcomes. For years, experts warned of the dangers of a pandemic and we are now living with it. For years, scientists have warned of a climate catastrophe that will forever change life on planet Earth. Now more than ever, nations and citizens must pay attention to the science and act on lessons learned. My book places faith in science and examines the role for science cooperation and the implementation of science diplomacy as a strategy to quell the rising tensions associated with contested sovereignty claims from Brunei, China, Indonesia, Malaysia, the Philippines, Taiwan, and Vietnam.

The South China Sea contains two major groups of features subject to overlapping sovereignty claims. The Paracel Islands in the northern half of the sea are currently controlled by China, but also claimed by Taiwan and Vietnam. The Spratly Islands, to the south, are far smaller, more dispersed, and even more contested. Both island groups, as well as a number of smaller features, have seen their share of violence, and the tangle of disputes appears intractable.

What distinguishes this book from others on the South China Sea is a hybrid of participatory research and field reportage. Since 2014, I have been a panelist and have organized a half dozen programs and podcasts with themes related to environmental security in the South China Sea with informed marine scientists and policy experts in attendance. Furthermore, I have been engaged as an in-the-field reporter for more than two decades in Southeast Asia and traveled aboard fishing boats, sampans and Vietnamese Coast Guard ships in the contested South China Sea.

I have known that fishers are the first to encounter the limits of the sea. This is most certainly true in the South China Sea, where overfishing in the region has emerged as a major threat to food security for populations bordering the churning sea. As a reporter and waterman, I have asked and answered central questions surrounding the patterns of overexploitation of the biological resources in the region, and believe that the environmental perspective can no longer be ignored. For sure, the historical, political and economic territorial claims in the region are mired in a complex and tangled web of nationalism. Perhaps the best marker for the future of these atolls, islands and rocks is examining the country or countries that are most equipped and qualified to sustainably develop and protect the islands' resources and diverse marine ecosystems. The claimant nations actions pose the natural question about how these patterns either are fostered or constrained by domestic, regional or international environmental politics?

In short, this book explains how conflict and cooperation can co-exist and that competing nations can, through environmental collaboration, adopt trust and science-driven peacebuilding measures that can influence and guide policy. Since the UN Decade of Ocean Science for Sustainable Development formally began in 2021 and calls for a new stakeholder process that will be inclusive, participatory, and global to deliver the science required for meeting sustainable development goals, this writing effort offers recommendations on how we can possibly build new relationships with non-science stakeholders and to embrace a new era of innovation, data sharing and scientific co-creation. The argument is simple: The South China Sea can become a body of water that unites; rather than divides.

The book's division into three parts—field notes, ecological politics, and science diplomacy—will lead the reader to understand that there is common ground that may lead to policy transformation and even resolution of maritime conflicts.

I hope that my introduction will draw you into a sustainable narrative by offering extensive anecdotes, field notes, and unvarnished conversations with fishermen and marine scientists about their responses to the resources in the South China Sea, living and non-living. As a non-scientist, I have always possessed a reporter's curiosity and have sought some modeled imitation of the observed reality. Fortunately, many generous biologists have allowed me the opportunity to see their seascape through a scientific lens. There's marine scientist, Dr. Chu Manh Trinh, a 59- year-old Da Nang University biology professor, who is responsible for mapping out protected areas off of Vietnam's central coast and for educating local fishermen about conservation, and how sustainable practices improve livelihoods in coastal and island communities. Professor John McManus, an American marine biologist and ecologist, has been diving deep into these roiling waters, studying the reefs and fisheries of the region. McManus along with other scientists, have observed the death of too many coral reefs from dredgers and giant clam fishers, which systemically tear up coral to capture their prize rainbow-colored ornamental shells that can reach almost 5 feet across and weigh in excess of 500 pounds.

The conservation and sustainability of fish stocks and maintaining ecological balance of the region require an understanding of how living and non-living South China Sea resources are rapidly being exploited by the people of the region. It's generally acknowledged that overfishing now threatens the South China Sea fisheries. That's one primary reason that since 1992, McManus and others have championed a clarion call for an international marine peace park. Of course, at the time, the islands were little more than atolls and rocks incapable of maintaining human habitation. He and other scientists, including Chinese marine scientists, have documented how the competition of natural resources, especially fish, have escalated the maritime claims of China, Taiwan, the Philippines, Vietnam, Malaysia, and Brunei in the contested Paracel and Spratly Islands. Sadly, all of these nations have viewed the sea as a resource prize.

Taiwanese Shao Kwang-Tsao, has also studied the coral cathedrals in the contested waters of the Spratly Islands, that prompted Taiwan's former President Chen Shui-bian to engage marine scientists through a proposed Spratly Initiative, and to recognize the region as an ecological protection area. Professor Shao, along with other dedicated marine scientists, conducted research around the area of Taiping Island, the largest of the naturally occurring Spratly Islands in the South China Sea, with its elliptical area of 0.87 mile in length, 0.25 mile in width, and an overall area of 110 acres. On this small island, the flora and fauna run riot: swallows, papaya plant, coast oak, lotus leaf tung tree, sea lemon, long stem chrysanthemum, coconut tree, banana tree, white-tailed tropic bird, sparrow hawk, and many kinds of tropical fish. The locale offers a proven dwelling for green sea turtles migrating from the Philippines.

But for these scientists, witnessing the gradual dismantling of life, beauty, and diversity in the South China Sea is a harrowing experience. Through their collective knowledge, the message is clear and certain: we are all in this together and the body of ecological science reminds us that life is interconnected.

The voices of these marine scientists and the fishermen remind us that we must protect all the parts of the systems on which we depend—from the smallest ecosystem to the hermit crabs I have seen on Cu Lao Cham. Rachel Carson's seminal, *The Sea Around Us,* a book that I have taught to Coastal Carolina University marine science students, reminds us that we must be faithful stewards of the sea. The scientist's poignant and prescient words spill over into our consciousness. She wrote,

> The little crab alone with the sea became a symbol that stood for life itself—for the delicate, destructive, yet incredibly vital force that somehow holds its place amid the harsh realities of the inorganic world [...] Underlying the beauty of the spectacle there is meaning and significance.

The facts are disturbing for all: in the last 50 years, half of the coral reefs have disappeared, only 10 percent of large fish remain and many species are at the brink of collapse. Unsustainable fishing practices, pollution—including 20 million tons of plastic entering the oceans yearly—and rising temperatures are continued threats.

The South China Sea's ecological collapse has been overshadowed by the current Covid-19 pandemic. We are still very much in the middle of this crisis and any real perspective on when it will end is near impossible to forecast. The chasm between the United States and China grows deeper and wider. The fiercest critics of China within the White House rail against the rule of the Chinese Communist Party, arguing that Beijing cannot be trusted to disclose what all it knows about the virus and especially how the outbreak began in Wuhan. Amid the escalating political tensions, staggering numbers of pandemic deaths, are now reaching in excess of 2.5 million global deaths and more than 520,000 in the US With quarantines and rising number of reported cases, and an untold surge of human suffering, the pain continues to be felt all over the world. That's the reason that an unprovoked sinking of Vietnamese Captain Tran Hong Tho's wooden trawler by a Chinese coast guard vessel in the disputed sea was lost in the storm of the current pandemic. While Tho and his crew of eight survived, there's little time for the world to pay attention to this wreckage against the current background of the global viral threat of the novel coronavirus and the mounting grim statistics of ravaged lives, shuttered stores, schools, and factories everywhere.

It was in the early morning with a red sky on the horizon, that a 33-year old Quang Ngai province fisherman said his goodbye to family and set sail with his crew to earn their livelihood near the Paracel Islands. It's a scene played out every day as these sentinels of the sea raise and lower their nets.

Through my field notes, fishermen share their voices about the sea and the decline of fisheries.

The tide turns daily from overfishing, destruction of coral reefs and deaths of fragile ecosystems. These fights over fishing rights alone represent a little-reported economic and environmental facet of the clash between China and its Southeast Asian neighbors in one of the world's hottest geopolitical trouble spots. But at the center of this sea of opportunities, uncertainties, and threats, environmental degradation, and death remain the scientific conversations. An increasing number of marine scientists, many of them personal friends and fishermen, are all sounding the alarm to address the issues of ocean acidification, loss of biodiversity, climate change and fishery collapse.

The book's middle section on ecological politics serves up the environmental details of Beijing's bullying and brutish behavior in the playground they prefer to call the South China Sea. As China's fishermen scrape every fish from the bottom of the roiling sea in their massive steel-hulled trawlers, destroy coral reefs, and ram competing nations' fishing boats, their offensive actions have immediate and far-reaching impacts. The ongoing fishing vessel skirmishes and continuing provocative Chinese-led actions, have cast long shadows and resulted in economic losses for many fishermen with boats sunk and equipment stolen.

There are an increasing number of challenges facing the sea and its inhabitants, but fish form the backbone of the powerful stories shared in my book. The book's narrative brings to life the challenges of food security from the perspectives of fishers, and marine scientists where the decline of fish is fast becoming a hardscrabble reality for more than just fishermen. In my conversations and reporting among all of them, I have learned about the dwindling fisheries in the region's coastal areas, fishing state subsidies, overlapping exclusive economic zone (EEZ) claims, and mega-commercial fishing trawlers competing in a multi-billion-dollar industry.

The sea offers very few natural partitions. My stories from the Mekong Delta to Ly Son Island, emphasize the interdependence of ocean and coastal land, life and water, atmospheric and oceanic circulation. For the Vietnamese, their identity is closely interwoven with their relationship to the East Sea and especially towards their fishermen. It holds them in a net of community, culture and heritage.

Finally, the book's promise is that scientific cooperation with regional authorities offers an important first step in building trust and confidence among neighbors and in implementing a shared conservation policy. It is time for more citizens and policy strategists to rally around marine scientists so that they can net regional cooperation and ocean stewardship to preserve them before it is too late. Science and diplomacy can and do work together. Science is neither political nor ideological. It is a universal language, because it promotes collaboration and openness.

While diplomacy is characterized by dialogue and negotiation, together, science diplomacy unfolds as a tool for "soft power" as characterized by Joseph S. Nye Jr., the distinguished Professor Emeritus and former Dean of Harvard's Kennedy School of Government. While the concept remains a new paradigm, more policy experts concede it contributes to coalition-building and conflict resolution sorely needed in the turbulent South China Sea.

What have I learned and what do I intend to share with readers? It's that the sea's complex and interconnected ecosystems need the voices of not only marine scientists but the families and fishers to quell the degradation wrought by such island reclamation, as well as overfishing and the harvesting of critical species that mars the region. I want to invite you aboard a traditional colorful wooden Vietnamese trawler, where a bone-tired crew, many as old as the leaking barnacle covered vessel, are too often the first casualties, along with truth, in the contested South China Sea. Let me be clear this is not another war story from Southeast Asia. The good news is that more marine scientists, some Chinese included, no longer want to see nations battling one another over these rocks and islands. They believe that it's not too late to chart a navigable policy shift for a sustainable and peaceful future.

Part I
Field Notes

1

The Gathering Storm

Many stories for journalists often begin in bars in Southeast Asia and this one is no different. In 2014, I was seated in a Hanoi café, drinking the locally brewed beer, *bia hoi*, with celebrated Vietnamese writer La Thanh Tung. It was a scorching summer afternoon. Nearby, the Song Hong river swept by. It flows irregularly because of the large amount of silt it carries. Rising in China, its deep and narrow gorge broadens into the river that feeds the fertile, densely populated Vietnamese delta. On a map, the river ends in Vietnam, and the South China Sea begins.

Tung speaks passionately about that East Sea, and the Vietnamese fishermen who depend upon it to feed their families. Men have fished there since before recorded history. Pollution, overfishing, and a dozen violent wars have diminished it, but still the boats return lying heavy in the water, hulls filled with tuna, mackerel, croaker, and shrimp. 50 percent of the animal protein consumed in Southeast Asia comes from the huge ocean shelf off the Vietnamese coast. He slowly sips his beer and says, "I think the sea is large enough for fishermen to earn their livelihood, but there are more challenges for them."

Two weeks earlier, a mammoth Chinese oil rig was parked in the middle of Vietnam's traditional fishing grounds. The international media—of which I am a member—is everywhere, its lenses focused on the maritime drama unfolding for Vietnamese fishermen. I learned that the Chinese occupied Vietnam for over a thousand years. They called it Annam, which means "pacified south." The protests and rioting on the coast, the Chinese warships protecting the rig, and my very presence here—confirm the irony in the long-forgotten meaning of that name.

I spoke with Tung about my recent interview with a fisherman in Da Nang (known as "China Beach" by the Americans during the Vietnam War), and one of Vietnam's major port cities. While there, I met Dang Van Nhan, a third-generation local fishing boat captain, who has been casting his long-line nets into the turbulent South China Sea for two decades.

Nhan recalled that on May 26, 2014, dark political clouds suddenly came over his horizon, and his weathered blue painted plank-constructed 50-foot fishing trawler was rammed and sunk by a Chinese naval vessel. As I was recalling my interview with the captain, I could still see the fisherman's plaintive eyes asking when do these attacks end since we are fishermen and the sea is open.

"The Chinese are plundering the sea with their huge trawlers and this hurts our ability to find the fish like we used to," says Nhan.

China's imposed annual seasonal fishing ban exacerbates the tensions for all fishermen in the contested waters. Beijing first announced this ban in 1999, broadly announcing that it would help sustain fishing resources in one of the world's biggest fishing grounds. The ban historically runs from May 1 to August 16.

The competition for fishing rights is one of the main motivations for the dispute over the waters, and observers warn that the Covid-19 pandemic could prompt a food crisis that would heighten the risk of conflict in the region.

The South China Sea encompasses 1.4 million square miles. It is of critical economic, military, and environmental significance. Over $5.3 trillion in international trade plies its waters annually. The region is richer in biodiversity than nearly any other marine ecosystem on the planet, and the fish provide food and jobs for millions of people in the ten surrounding countries: Brunei, Cambodia, China, Indonesia, Malaysia, the Philippines, Singapore, Taiwan, Thailand, and Vietnam.

In the South China Sea there are approximately 180 features above water at high tide. These rocks, shoals, sandbanks, reefs and cays, plus unnamed shoals, and submerged features are distributed among four geographically different areas of the sea. In turn, these aspects are claimed in whole or in part by China, Taiwan, Vietnam, the Philippines, Malaysia, and Brunei.

My story was making me thirsty, and so I had another draft beer and described as many details as I remembered about Nhan's boat since I viewed it after it was first pulled out of the roiling sea and brought into Da Nang's boat yard, where I stood with the forlorn captain examining the wreckage.

The damage was evident when I first climbed an improvised wooden ladder to closely inspect Captain Nhan's boat. It had been salvaged following its sinking and was now on display for foreign media in one of Da Nang's boat yards. The deep gouge inflicted by the steel-hulled Chinese boat on the Vietnamese fishing boat's starboard side revealed the mortal wound that had sunk the once buoyant and colorful traditional fishing vessel.

The hull shape is long and slender, with a high overhanging sharp bow directly derived from the country's traditional sailing vessels, but the stern is not pointed and runs to form a flat transom. The engine is placed directly under the wheelhouse. The fishing appendages are fitted on the foredeck, where sometimes the remnants of a mast serve to hang up lights for squid fishing or a protective tarpaulin. The bow is always festooned with a traditional eye or "oculi" and with a *nga* (cathead). These eyes are long and painted on both sides, distant from the stem or median plank.

Unfortunately, for this fisherman and crew on this fateful morning, their traditional and ceremonial decorative eye failed to protect them from a Chinese vessel that deliberately smashed their livelihoods.

According to Charlotte Pham, an Australian scholar on Vietnamese boats, "Some hulls in a northern tradition have plank edge-joined by nails, on a keel plank that rises fore and aft, and ends with a flat transom. Hull planks in the center and south of Vietnam are edge-joined with dowels and have high sterns and prows overhanging the water. Other hull planks were stitched, and some bamboo rafts were fitted with sails and daggerboards."[1]

Nhan's boat was certainly no match for the mega-ton steel-hulled Chinese boat. Over the last decade, Xi Jinping has promoted the maritime industry and urged mariners to "build bigger ships and venture even farther and catch bigger fish." Some experts claim that China's distant water fishing fleet numbers around 2,500 ships, but one study claimed it could have as many as 17,000 boats trawling and plundering the oceans.[2]

The United Nations confirms that China is a fisheries superpower since it consumes around 36 percent of total fish production, and hauls in more than 15.2 million tons of marine life a year, or 20 percent of the world's entire fish catch.

Tung ordered me another beer since he wanted to hear more of the fisherman's story.

The 56-year-old boat captain and other fishermen, like Bui Ngoc Thanh, have always been aware of the perils of a seaman's life. Squalls capable of upending a trawler spring up quickly and a fast-moving typhoon can easily outrun a ship. Their lives have always been at risk, but this monolithic rig and its heavily armed Chinese escorts posed a new and unfamiliar threat to their livelihood.

For sure, the new China has made steady advances over the past two decades to reclaim its role as the pre-eminent power in the Pacific. Because of their investments into the maritime industry, what has emerged is the rising level of competition and even conflict in the South and East China Seas, driven by fear of losing control of key supply lines, competing maritime claims, differing interpretations of maritime agreements, and competition for resources, especially fish.

China Dialogue, an independent, non-profit organization dedicated to promoting a common understanding of China's environmental challenges, states that China is now among the world's largest ocean powers. It has the largest fishing fleet, emits the greatest volume of greenhouse gases, produces one third of the world's ocean plastic pollution, maintains the world's biggest aquaculture industry, and harvests and consumes more of the world's seafood than any other nation.[3]

With their thousands of steel-hulled mega trawlers operating far outside their exclusive economic zone (EEZ) spanning from the South China Sea, the Pacific, the Federated States of Micronesia, Papua New Guinea and to the African coasts, it's no wonder that China is recognized as the major driver of illegal, unreported, and undocumented (IUU) fishing globally.

As the afternoon peacefully slipped away at the café with the nearby Tamarind trees offering little respite from the heat and humidity, I decided to digress, like the Southerner that I am, to ask Tung some sensitive questions. Since he is a writer, I knew that he was expecting them. After all, he has seen much in Vietnam's fast-moving renovations post the American War. But my intention was not to speak about a war that cost 58,200 American lives and more than 1.1 million North Vietnamese and Viet Cong fighters. I wanted to better understand Chinese and Vietnamese relations that span two thousand years.

"My country has lived with Chinese threats for more than my lifetime. We clashed with them in a brief war in 1979 in the mountainous northern border in Lao Cai and in a naval battle in 1988, in the Spratlys, where 64 Vietnamese soldiers died," sighed Tung.

I knew these were sensitive subjects for the two countries are ideological brothers who share a belief in communism, an ideology largely abandoned by much of the world. That helps explain why the 1979 border war is something not too many Vietnamese want to speak about. The other equally delicate topic are the territorial claims on The Paracel Islands, called Xisha Islands in Chinese and Hoang Sa Islands in Vietnamese. They lie in the South China Sea approximately equidistant from the coastlines of the PRC and of Vietnam at nearly 200 nautical miles. With no native population, the archipelago's ownership has been in dispute since the early 10th century.

"Of course, our relationship with China is complex. We honor the Chinese philosopher Confucius at our Temple of Literature right here in Hanoi. But how can we forget the 1979 border war when Chinese troops advanced on our citizens in the province of Lang Son," interjected Tung.

I joked that that the South China Sea documents the transition from wide continental rifts to narrow rifts in geological time dating back to the Eocene. Since I am a non-scientist, Tung laughed when I added that maybe that the tectonic shifts help explain the China and Vietnam history.

This brief war between Hanoi and Beijing on the northern border was attributed to several factors, including to punish Vietnam for its invasion of Cambodia in the early winter of 1979 to oust the Chinese-backed Khmer Rouge. Beijing sent more than 80,000 Chinese troops across the border.

As Tung chain-smoked his cheap Vietnamese cigarettes, he was quick to interject, as if he were already anticipating my question about the 1988 Sino-Vietnamese clash at Johnson Reef or *Gac Ma*. This event memorialized in Vietnam, occurred when a Chinese naval frigate sank two Vietnamese ships, leaving 64 sailors dead, some shot while standing on a reef, and to this day it remains a serious inflection point between the two nations.

The aftermath of that skirmish is felt today by fishermen in the contested South China Sea. China secured its first Spratly Islands foothold where the early fortifications remain critical markers in the geopolitical tensions between Beijing and Hanoi.

Tung quietly shared details of that battle. He looked around the bar to see if any eyes were on us or rather on him as he narrated his understanding of the very political sensitive historic battle that took place on March 14, 1988, when 64 Vietnamese fishers and soldiers lost their lives protecting a reef in the Spratly Islands.

The scene ensued in the early hours of a red sky morning, when Chinese warships first approached the South Johnson Reef, nearly submerged and encircled by white coral reefs and located in the middle of the Spratlys. It was reported that the Vietnamese sent several small boats with crew to take possession of the reef. About a half dozen Vietnamese soldiers bravely waded into the waters to attempt to plant their nation's flag on the rock. The Chinese wasted no time and fired on the men. It was no match. The Chinese warship fired its canons at the remaining Vietnamese aboard their boat, and it sank.[4]

32 years later, the tragedy remains memorialized. *Gac Ma* is now claimed by China, Taiwan, the Philippines, and Vietnam. In the intervening years, China's power has dramatically increased as the second largest economy in the world, and even in the midst of the present pandemic, it is expected to surpass the US to become the biggest economy in 2030.

Although the state-controlled media sometimes publishes articles on the topic, with headlines like, "Officials award the Vietnam flag for fishermen to encourage fishing." The political sensitivity surrounding the *Gac Ma* attack has not gone away.

In email communication with the book publisher First News based in Saigon (Ho Chi Minh City) that was tasked with publishing the book *Gac Ma The Immortal Circle*, I was informed that upon publication (Vietnamese language edition) that the book was recalled because of one specific passage cited by a survivor of the battle.

"I intended to pick up the Chinese captain's gun and shoot him dead, but since there was an order of not to shoot so I didn't. If allowed, I would kill him! In the end, I was both stabbed and shot."

I want to protect the Vietnamese source at First News, since he revealed to me, that the publication of the words, "not to shoot," defamed the Vietnamese Army, as cowards towards the enemy. Although before the book was published, there were many blogs and commentaries about this order and so it confirms the truth of the incident. After the criticism, the book publisher fixed it and changed the phrase into "not to shoot first," but this proved insufficient and the book was withdrawn from publication. This was after more than four years spent researching and interviewing nearly 68 participants, including generals, historians, journalists, and Ga Ma veterans in the completion of the book.[5]

Since China remains a leading trading partner with Vietnam, Hanoi carefully and cautiously enables, or and at times, orchestrates its citizens to mobilize anti-China protests. It seems that China and Vietnam would be the most natural allies since they both share a common belief in communism and have adopted a market-driven capitalism that has transformed both countries in so many ways. Never mind that the two countries' complicated history of territorial disputes has at times ruptured geopolitical ties, President Xi Jinping has more than once referred to Sino-Vietnamese relations as a "special friendship between comrades and brothers."[6]

But the Vietnamese respect and national identification for their East Sea (they prefer this name to the South China Sea) often does spill over into streets and has resulted in damage inflicted on Chinese stores and factories. What is clear is that the Vietnamese will not stand for the ongoing harassment and ramming of fishing vessels, the confiscation of their fishermen's equipment, and death and injuries inflicted upon their fishermen.

"The Spratly and Paracel [islands] are still occupied. Of course, we must attempt to put our differences aside. But we want our islands back since they are our ancestral places, especially for the fishermen," added Tung.

2

Bad Luck Sinks Fisherman's Dreams in the Contested Sea

As a fisherman, 50-year-old Bui Quang Mong has faced many challenges casting his frayed nets into the South China Sea. He's stared down typhoons, been chased and caught by Chinese navy ships, fought over confiscated catches, been detained, and most recently, forfeited his boat to a Vietnamese bank.

Mr. Mong's bad luck or *xui xeo* has followed him since 2008, when his first wife died from cancer, leaving him with two children. His second wife, 36 years old, has been diagnosed with thyroid cancer, and has just undergone her fourth round of radiation therapy. They have a son, who is in first grade, living with his mother.

Fishing did not run in the family. His dad came from a hardscrabble farming life in Quang Ngai province in Vietnam's Central Coast without any experience in seafaring. In 1992, after completing his military service, Mong decided to become a fisherman. His first job was working as cook aboard a squid fishing boat on the Central Coast. During his time working as cook, he started to learn engine repair from the ship's mechanics.

The ship owner's Sat Mot, (his real name is Sat, but since he had only one eye, people called him "Sat Mot"). Sat used his foot to kick Mr. Mong in the head as his daily wake-up call, and who too often behaved like a brutish Captain Ahab, the complex antihero of Herman Melville's *Moby-Dick*. Once he was kicked, Mr. Mong got up to go to the toilet when the boat owner yelled: "I demanded that you get up but not to pee." After that, every time he was kicked, Mr. Mong went quickly into the engine room to check the grease, then raced to the galley to make Vietnamese coffee for his tyrant boss and prepare meals for crew members. Working as a cook, he was at the mercy of the captain and the fishermen on board the trawler.

Fortunately for him, he was a quick study and began to learn everything about the engine. Every time the engine of the ship breaks down and this was a frequent occurrence, Mr. Mong was responsible for holding the light so the mechanic could fix the engine. Because he wanted to become a mechanic, the tireless seaman noticed every detail of the mechanic's movements and kept asking so many questions that the mechanic also became irritated. Mr. Mong said that he wanted to learn how to repair the ship's engine so that he could fix it during seafaring trips, when there was no mechanic around. Finally, the mechanic consented to teach the cook about each part of the engine, each type of failure, and how to fix them. Gradually, he gained experience, earning the trust of the shipowner to repair the ship's engine and even promoted his engine repairing skill to other ships. After more than a year of working, Mr. Mong was promoted from his position of cook to a "friend" or *Ban ngu dan*.

The boat mechanic's brothers fished near the shore and were able to make a decent living with stable families. But Mr. Mong had bigger dreams, and with those dreams came greater risks. He eventually borrowed money from a bank to go fishing offshore in a custom-built composite-constructed trawler into the East Sea, what the Vietnamese refer to as the South China Sea.

This body of water encompasses an area of around 3.5 million square kilometers and is formed from the marine, coastal, and river catchments of ten nations: China, Vietnam, Cambodia, Thailand, Malaysia, Singapore, Indonesia, Brunei, the Philippines, and Taiwan. In Southeast Asia, where marine biodiversity is particularly high, over 300 million people live within the coastal zone. The South China Sea consists of over 250 atolls, cays, islands, shoals, reefs, and seamounts.

According to the United Nations Environment Program, the region provides some of the world's most diverse sea grass beds and mangrove forests, as well as more than 2,500 species of marine fishes and 500 species of reef-building corals. The fisheries in the South China Sea are of significant local, national, and international importance since they are major contributors to the food security and economy of the bordering countries.

From trawlers, some steel-hulled, some composite but mainly wooden and mostly leaking, fishermen are competing for tuna, mackerel, flying fish, billfish, and sharks for the pelagic species along with a large array of demersal fish and invertebrates, especially penaeid shrimps.

The catches of pelagic fish species are estimated to be about 3.9 million tons (accounting for 45.3 percent of the total landings), according to a study in *Nature*.[1] These highly migratory fish species are generally recognized as swimming across the EEZs of more than one country and international waters, and thus are also known as shared stocks. However, Vietnam's neighbor China has little interest in sharing their fish. Beijing's turn to the seas and its declaration to become at all costs a maritime power and a respected member of the international community have brought much pressure on the fishing industry.

Under the driving leadership of Xi Jinping, China has vigorously staked its claim to most of the South China Sea as its sovereign territory within the so-called "nine-dash line." The rising power has been encroaching on territory within the 200-mile exclusive economic zones (EEZ) of Vietnam, the Philippines, Malaysia, and Indonesia; threatening force against US military intelligence, surveillance, and reconnaissance (ISR) activities in China's EEZ; and protesting US Freedom of Seas Naval Operations (FONOPs).

Beijing's urge to control the seas is all part of the nation's development and for achieving the China Dream. This is the lens that fishermen are using as they drive their fishing vessels into the dangerous geopolitical waters. By some calculations and with detailed reporting from Ian Urbina of *The New York Times*, the Chinese government places its distant water fishing fleet roughly at 2,600 but other sources, including a study by the Overseas Development Institute (ODI), puts this number at 17,000.

Mr. Mong and other fishers in the region, need no further education about how this sea is one of the world's most contentious areas with significant territorial disputes among neighboring countries. There have been territorial disputes between China and Vietnam over the sovereignty of the Paracel Islands, which have been occupied by China instead of Vietnam since 1974. Among China and Taiwan and their Southeast Asian neighbors, disputes endure over the sovereignty of the Spratly Islands and other offshore resources.

Although Vietnamese authorities attempt to forecast fish stocks in all fishing grounds, Mr. Mong thinks that it is inaccurate because the forecast is based on fishermen's voyage diaries, which are often not recorded or patently false. Fishermen generally keep their secrets closely guarded about where they can find abundant fish stocks. The universal fishing aphorism is that there are two ways to tell the experience of an angler, how he holds a fish and how he keeps his secrets. The latter, especially among Vietnamese, is always more important.

As a result, fishermen often rely on experience to find fishing sites. However, the weather also plays an important role in their success and failure. The storms are cruel. In fact, almost 15 cyclones, typhoons, and storms occur annually in the churning sea and claim hundreds of lives. However, the dangers of the sea do not stop fishermen from sailing out of their safe harbors.

The fishing volume started to decline when China encouraged their fishermen to venture further into the South China Sea. Then, in the midst of this fishing frenzy, Vietnamese authorities began closely monitoring the navigation of fishermen in an effort to remove the European Union "yellow card" designation over illegal, unreported, and unregulated (IUU) fishing. The plundering of the sea was aided by Chinese ships that are much bigger, have a larger holding capacity, and are equipped with many fish-finding electronics, so Vietnamese ships cannot compete or are at a major competitive disadvantage. For example, when using lights for fishing, Vietnamese ships usually only have about 40 lights, but Chinese ships use up to 500 lights. In addition, Chinese ships also sneak into Vietnamese waters at night to catch fish. As a result, seafood resources are becoming increasingly exhausted.

As a result, it often does not pay for fishermen to go too far beyond 200 miles out because a trip's revenue cannot cover expenses or break even. Mr. Mong, who has never forgotten his difficult life as a farmer in Quang Ngai, wants the government to participate to solve problems for fishermen. He also said that so many others are in similar circumstances. He was able to apply for a commercial boat construction loan under the Vietnamese government's Directive No. 67, designed to facilitate low-interest loans to assist qualified fishermen to build new boats or upgrade existing boats.

He approached the innovative Institute of Ship Research and Development (UNISHIP) in Khanh Hoa province on Vietnam's central coast. In association with the University of Nha Trang, the small shipyard has launched several fiberglass vessels in 2020. Having demonstrated to the bank that he was both deserving of a boat loan and capable of repaying it, he was able to order a new 24 meters by 6.50 meters beam glass reinforced plastic (GRP) fishing boat, named *Ju Mong Truong Sa*.

During the loan application process, he was asked why a GRP boat and a relatively expensive engine were on his list of requirements, and his response was that he was looking for reliability and low maintenance costs that would allow the boat to fish more effectively, allowing him to repay the loan more quickly. But he may have been reflecting that these waters are as much Vietnamese as Chinese, and with a modern boat he believed that he can not only make a living for his family and crew but also help demonstrate his nation's claims to sovereignty.

In our interview, the Vietnamese fisherman stated that "he wanted to build a fiberglass boat with a big expensive engine, since it would enable him and his crew to travel further and faster." Yes, his declaration was clear and true. The boat was needed to follow the fish and a nation's sovereignty claims.[2]

In June 2016, a proud and determined Bui Mong took delivery of his new fishing vessel with a modeled depth of 3.5 meters and a 10,000-liter fuel capacity. With a packing capacity of 70 cubic meters in eight compartments, it allowed efficient use of ice and provided storage for significant catches of tuna and other fish species. Mong got his powerful and reliable engine, a 12-cylinder Cummins KTA38 engine delivering 800 HP at 1800 RPM.

His fishing vessel was 95 percent funded by the state, and he contributed the remaining 5 percent. Its engine already cost 3 billion VND (around 130,000 USD), and the total value of the ship was 12.5 billion VND (around 538,000 USD). In comparison, a traditional wooden fishing boat of good quality costs only 1 billion VND (around 43,000 USD).

Those fishermen in Vietnam's central coast who took out loans from commercial banks to buy steel fishing boats under a government-backed scheme have reported huge losses. At least ten owners of steel-hulled fishing boats signed a petition to complain about the quality of vessels built by Nam Trieu and Dai Nguyen Duong Shipbuilding Company, claims Vo Dinh Tam, a senior fishery official from Binh Dinh province.

From the traditional coracle bamboo basket woven boats to the multi-colored traditional blue hulled wood trawlers, Vietnamese fishermen in each port, and coastal fishing town, are busy repairing their nets in anticipation of another day casting for their living. According to scholar Charlotte Pham, she believes that the boat building traditions of Vietnam require much more extensive research because of the diversity of boats. Her doctoral dissertation offers many historical references and documentation on over 50 types of Vietnamese boats by size, construction materials, fastenings and engines.[3]

When I was in Cu Lao Cham, a pristine eco-friendly island located about 12 nautical miles from Hoi An, I watched old fishermen, men and women, take their little round bamboo basket boats and skim it effortlessly out of the harbor with the use of an oar. However, Vietnam's traditional wooden trawlers are little match for China's steel-hulled fleet. There's growing competition for the dwindling fish in the disputed South China Sea. According to the Washington-based Center for Strategic and International Studies (CSIS), the sea accounted for 12 percent of the global fish catch just five years ago. However, there remains a lack of transparency in the collection of data.

What is clear is that the South China Sea has been dangerously overfished and catch rates have declined by 70 percent over the last 20 years. Now more fishermen have fully realized that bigger and faster boats only translate as less fish to catch.

Yet Mong felt confident in his decision to have a composite trawler built, since too many modern steel constructed boats that cost hundreds of thousands of dollars were foundering only a few months after they were launched generally from inferior steel purchased from Chinese suppliers. In fact, too many fishermen in Vietnam had issues with their steel hulled boats from the shoddy construction and inferior steel that rusted quickly.

"The steel is from China and is of a poor quality and this only leads to damage and disappointment for the fisherman," claims Mong. He suspected that shipbuilding companies' corruption is responsible for damaged ships and that some companies bought the inferior steel in order to receive more funds from the state budget to compete against the Chinese steel-hulled trawlers.

Unfortunately, one fisherman, an acquaintance of Mr. Mong, built a steel-hulled ship, also under Program 67. However, his ship was heavier, so it couldn't chase the fish and returned to port with many problems. The steel-hulled ship's ice cellar was too hot, so the ice melted quickly, decreasing the quality of the fish. As a result, the fish could only be sold at a lower price. Moreover, these steel-hulled ships deteriorate after 3 to 5 years and each repair also costs a lot of money.

When his wife became ill, Mr. Mong could not go on seafaring trips and chose to stay at home to take care of her, and the composite vessel was left in the port. But he still was faced with the demands to pay salary for his crew as well as the government loan to build the vessel (a 16-year payment).

The bank placed him in the list of individuals unable to repay debt and sued him in court. He was required to pay the court fee of more than 100 million VND (around 4,300 USD). He refused to go to court, the court still tried him in absentia and sided with the bank, confiscating his ship.

He was forced to say goodbye to his seafaring life, although he did not want to. He changed to another job and left behind a few friends who were also former fishermen. He now transports wooden furniture from the Central Highlands to Dong Ky village in Bac Ninh Province that are for sale.

At home now, Bui Quang Mong misses seafaring most days. The decline in fishing grounds, along with more Chinese fishing vessels on the horizon, only serve to cast a longer shadow over him and the sea.

3

After the Storm

The air was sultry and rain imminent when I flagged down a Hanoi motor-bike driver to learn more about the tug-of-war over the sea and its many islands. Some friends think me brave for jumping on the back of these motor-bikes since the pollution-choked capital has more than 5 million motorbikes on the clogged city streets—many of them carrying entire families.

I went to speak with an expert on the East Sea in Hanoi about China's efforts to place their oil rig in Vietnamese waters.

"The location of the oil rig is outside the Paracels Exclusive Economic Zone, and it has been placed there as a strategic weapon of foreign policy," claims Dr. Tran Troung Thuy, director of the Institute for East Sea Studies at the Diplomatic Academy of Vietnam in Hanoi.

To better understand the ongoing geopolitical security issues, I knew that I needed to dive deeper into the myriad of issues. The Global International Waters Assessment (GIWA) characterizes the South China Sea as severely over-fished, with excessive by-catch and discards, and destructive fishing practices, including cyanide, dynamite fishing, and gill nets.

Dr. Nguyen Long, Deputy Director of the Research Institute of Marine Fisheries (RIMF) believes that "overexploitation of coastal resources is demon-strated by a decline in fish catches with specific declines in lobsters, abalone, scallops and squid."

Growing disputes over fishing territories between China and other South-east Asian claimants have added to Chinese and Vietnamese frustrations over the current situation in the South China Sea. Accounting for around 10 percent of the world's annual fishing catch, this region has been a historical fishing ground for both nations' fishermen.

No one disputes that the South China Sea fisheries offers a cheap supply of food, a means of livelihood, and a source of foreign exchange. In addition, the sea's nutrient-rich waters provide the habitat and spawning grounds for the world's most valuable shrimp and tuna fisheries.

A large portion of the coastal workforce is dependent on the marine environment through employment in fishing, transport, recreation and tourism, offshore exploration, and extraction of hydrocarbon and other natural resources.

In recent years, China's assertive behavior and increasingly frequent conflicts with other claimant nations overfishing in the disputed area have caused diplomatic tensions and sometimes heightened mutual public hostility. China and Vietnam have adopted concerted media campaigns amidst claims that fishermen have engaged in illegal fishing in areas surrounding the Paracel Islands, resulting in attacks, injuries and imprisonment.

"Managing the fisheries of the South China Sea region is rendered immensely difficult by the non-cooperation of the various countries, which all compete by building their fishing effort. Moreover, the fishery catch statistics are largely driven by political imperatives, since they report ever-increasing catches, which have little to do with reality," says Dr. Daniel Pauly at the University of British Columbia Fisheries Centre.[1]

The provocative placement of China's oil rig deep into the Vietnamese continental shelf, and just 17 nautical miles from Tri Ton Island, demonstrates China's strategic aim of dominance over the region's fisheries. The positioning of four more Chinese oil rigs in the region adds measurable tension to the already tempestuous political waters.

The flash point issue is deliberately obscured by China's less than transparent territorial boundary: the "nine-dash line," first introduced in 1947 by the Chinese Nationalist government. It was then submitted by the People's Republic to the UN Commission on the Limits of the Continental Shelf in 2009, as a response to the Vietnamese-Malaysian joint submission for extending their maritime rights in the southern part of the South China Sea earlier that year.

China's aggressive mapping of the region includes a ten-dash map, underscoring its ambition to claim over 90 per cent of the South China Sea.

For that matter, there are well over 260 academic articles that examine China's adoption of the nine-dash-line and its collection of arbitrary dashes and dots to mark its claims to these historic waters. According to researcher, Nguyen Thuy Anh, a fellow at the East Sea Institute in Hanoi, these numerous map references, only succeed in "prompting misperceptions among scientists, researchers, or students. The concern is that policymakers in Beijing or others may view these publications as truth rather than propaganda."[2]

"Beijing has been careful not to clarify exactly what the line means because that ambiguity gives it significant room to maneuver. The best explanation, as laid out by Chinese legal experts and academics, is that the nine-dash line only lays claim to the islands and adjacent waters within it, but also entitles China to undefined 'historic rights' within all the remaining waters," according to Greg Poling, a scholar at the Center for Strategic and International Studies (CSIS) and the director of the Asia Maritime Transparency Initiative at CSIS. He oversees research on US foreign policy in the Asia Pacific, with a specific focus on the maritime domain and the countries of Southeast Asia.

Before China parked its giant oil rig, *Haiyang Shiyou 981*, in the disputed area, CNOCC's chairman Wang Yilin claimed at its 2012 launching that "large-scale deep-water rigs are our mobile national territory and a strategic weapon."[3]

More than 86 Chinese ships, including three warships, offer protection to the rig, and reinforce that bold statement. This coercive rhetoric directed against its regional neighbors serves notice that the Middle Kingdom intends to extend its sovereignty well into the South China Sea.

In Hanoi at the popular expat-populated Moca Café, near the historic St. Joseph's Church, I met with Tang Mengxiao, a doctoral candidate in Political Science and International Relations at the University of Southern California, we were discussing the deployment of the $1 billion Chinese deep-water oil rig at about 120 nautical miles off of Vietnam's coast, in what Vietnam considers its exclusive economic zone (EEZ). There's no disagreement that this has led to the worst breakdown in relations between the two countries since their brief border war in 1979.

"The oil rigs are a 'strategic weapon' and can only be perceived as China's way of declaring sovereignty, displaying power and exploring marine resources in the disputed South China Sea, especially in the wake of Obama's trip to the four Asian countries and the US gesture of renewed commitment to the region," says the graduate student.

The strategic importance of the contested sea dates back to when China attempted to seize the Paracels in 1974, but was unsuccessful in its 1982 assertion during the UN Convention on the Law of the Sea negotiations, that the archipelago was its territory.

Some experts on these territorial matters, like Dr. Martin Murphy of Georgetown University believes that "if China's blatant territorial moves go unchecked and its blue-water naval capability expands, it is likely that it will have the power to shift the international rules governing the maritime domain."[4]

No one disputes that the diplomatic hotlines between neighboring countries that share the waters of the world's largest commercial sea-lane remain heated. However, China shows no sign of backing down from the pressures imposed by the Philippines, Japan, Vietnam and even the US. Chinese coast guard vessels continue to ram Vietnamese fishing ships and attack them with water cannon.

In the summer 2014, I attended the two-day CSIS conference in Washington about the South China Sea. Chu Shulong, a professor of political science and international relations from Tsinghua University, adopted the Beijing party line and defensively shared with the many academics, policy shapers, and reporters present, the view that China always follows the road of peaceful development and never seeks confrontation. However, he was stridently critical of America.

He bluntly claimed that it does not make sense for the US to accuse China of threatening the use of force or coercion when it is increasing its own military presence in the South China Sea. "What should we call this? It's also threatening the use of force. It's also coercion," repeated Chu.

The drama at sea over access to oil and gas resources is coupled with both China and Vietnam's over-exploitation of the fisheries, since their fishermen use long lines and gill nets. Some Vietnamese fishing trawlers are trolling in off shore waters for oceanic or flying squid, angling, drift gill nets and drifting long lines. In speaking with Nha Trang fishermen, they confirmed that operate their drifting longlines for tuna and other pelagic species. Their long lines are used at nighttime and the catch are skipjack tuna, yellowfin tuna, and marlin.[5]

Certainly, China's appetite is at least four times that of its neighbors and with its fisheries in a state of near collapse, the Chinese authorities are poised to assert their expansionist plans to seize an even greater share of the declining regional fish stocks.

Nevertheless, China's Foreign Minister Wang Yi has taken the diplomatic high ground and insisted that "we are in this boat together with more than 190 other countries. So, we don't want to upset the boat, rather we want to work with other passengers to make sure this boat will sail forward steadily and in the right direction." This metaphor does nothing to diminish the Middle Kingdom's worsening coastal fishing crisis that propels their streamlines steel trawlers into deeper contested waters. In response, Vietnam's fishermen have been granted generous soft loans to replace their traditional wooden boats with steel hulls or fiberglass construction.

Excessive and unsustainable fishing practices, as well as land-based pollution, coral reef damage, and other factors, have exacerbated the depletion of fisheries. It's no surprise that policy experts are concerned about the implications of escalating tensions.

The food security and renewable fish resource challenges are clear. According to a World Bank Fisheries Outlook, "Fish to 2030: Prospects for Fisheries and Aquaculture," China will increasingly influence global fish markets. A baseline model projects that China will account for 38 percent of the global fish consumption by 2030.[6]

In fact, an increasing number of fishery experts express concern for the region's hard and soft corals, parrot fish, spinner dolphins, sea turtles, groupers, and black-tipped sharks. From Vietnam's coastal areas to Hainan Island, the region has experienced a 60 percent coral life and 50 percent fish species decline.

"Over the past 40 years, it is estimated that fishery resources in Southeast Asia have been reduced to 25 percent or less of their former levels," claims Dr. Ralph Emmers from the S. Rajaratnam School of International Studies at Nanyang Technological University.[7]

It's noteworthy that of the 3.2 million fishing vessels operating in marine waters worldwide, an estimated 1.7 million, or nearly 55 percent, are in the South China Sea. Most of them are small scale fishing boats. Since the skirmishes between Vietnamese fishermen and Chinese coast guard vessels show no sign of easing off, these incidents in disputed waters will continue to be a cause of diplomatic tension and to provoke greater nationalistic sentiments.

The continuing clashes reinforce the view that fish is a strategic commodity to be protected and defended, by force if deemed necessary. Alan Dupont and Christopher G. Baker at the Lowy Institute for International Policy in Sydney, Australia, argue that "Beijing is using its fishing and paramilitary fleets for geopolitical purposes in pursuing a strategy of 'fish, protect, contest, and occupy,' to enforce its sovereignty and resource claims over contested islands and coerce other claimants into compliance with, and acceptance of, China's position."[8]

In mid-July 2014, a Category 1 tropical storm barreled towards the Philippines. The storm was upgraded to a 2 and then a 3. The storm, at first named *Glenda*, was given an international name, *Rasmussen*. Bobbing out on the horizon, small fishing boats from Vietnam and the Philippines disappeared from view.

The boats are mostly crewed by men. Each deckhand is someone's son; each vessel's pilot is somebody's father. Someone's brother's hands will burn from pulling on a net woven by someone's mother.

At this time, the Chinese oil rig was 130 nautical miles from the coast of Vietnam and could not be seen from the shore. This oil rig and the deep waters beneath it had been in the quiet eye of an international storm for several months. *Rasmussen*, now a full-force typhoon, spared Manila much of its wrath, but left at least 50 dead. Several fishing boats and their crews were missing, presumed dead.

Rasmussen was soon classified a super-typhoon. Perhaps the most powerful storm in recorded history, it was then predicted to hit Southwest China, followed by a third landfall in Vietnam. But 33,000 feet above Ukraine, Malaysian Air Flight 17 was reportedly struck by a Russian built SA-11 missile. Close to 300 passengers were presumed dead. *Glenda*—or *Rasmussen*—is no longer the story.

Over 3,000 miles from the wreckage, and even further from the media coverage, China oil rig *HYSY 981* was leaving the South China Sea. Still protected by warships, powerful tugs were towing it safely away from the storm's path.

Almost exactly where the oil rig was first moored and all the surrounding fury began, the storm surrendered itself to the ocean. Downgraded quickly to a Category 1 storm, minimal damage was reported in China and Vietnam. No fishermen were reported lost or dead. The next morning, with no storm, rig or reporter in sight, the fishermen of Vietnam headed out onto a quiet sea.

The fishermen of Vietnam have had a good haul this season. The real threat of the rig is to their right to fish these waters freely. Perhaps even to fish them at all.

The rig will be back. But for now, as is always the case after a storm, the fishing is good.

4

A Perilous Passage[1]

They were called the "boat people." The exodus of refugees from Vietnam began even before the fall of Saigon on April 30, 1975 for South Vietnamese political leaders, army officers, and skilled professionals. Waves of refugees took to the treacherous South China Sea in makeshift, unseaworthy, and overcrowded fishing boats. Tens of thousands pushed off from the shores hoping that the oceans which had provided the generations before them with life, would provide this one last passage to safety.

On one of the many flights I have taken to and from Vietnam since the War ended, I was seated next to Dan Nguyen, a 64-year-old Vietnamese, who spoke to me about his harrowing passage from Vietnam in 1981 with his two younger brothers, aboard a leaky wooden fishing boat on the churning South China Sea. I wondered what it would take for me to brave piracy, the unforgiving sea, and the darkness of countless unknown dangers. How desperate would I have to be? What promise and hope would a foreign place have to offer?

Nguyen and his brothers survived. More than that, they succeeded beyond mere survival. Arriving in California, Nguyen went to work for Cisco Systems, an American multinational conglomerate headquartered in San Jose. As more refugees arrived on the California coast, it was as if these refugees took a page straight out of John Steinbeck's novel, *The Grapes of Wrath*, where all things seemed possible and where they could rebuild a community of family.

Mr. Nguyen shared with me on our long flight how he established a successful career with Cisco, raised a family, and became an American citizen.

Nguyen's story along with more than 130, 000 other lives are all unique, and so many different versions are recounted in Thurston Clarke's historical book, *Honorable Exit: How A Few Brave Americans Risked All To Save Our Vietnamese Allies At The End of the War*. The narrative reaffirms humanity's need to pull together. On that long flight home, I learned more from my fellow passenger about how people, especially refugees, depend on one another. The events that followed the fall of Saigon mark the start of the refugee crisis and exodus of millions of people fleeing Vietnam.

The United Nations High Commission for Refugees estimates that between 200,000 and 400,000 Vietnamese refugees died at sea between 1979 and 1986. 800,000, like Dan Nguyen, survived.[2]

Between 1975 and 1994, the Vietnamese diaspora reached almost 1.4 million across the world.[3]

Although not classified legally as "refugees" under international law, Vietnamese were commonly called refugees as well as "evacuees" and "parolees." Many of the Vietnamese who fled sought and soon gained resettlement in the US. President Gerald Ford's administration allowed them to enter as "parolees;" a loophole in the US immigration policy, which did not make provisions for refugees at that time.

For the war evacuees, the escape from Vietnam was triggered as a result of the communist government of North Vietnam, supported by the Soviet Union and China in a battle with South Vietnam, backed by the US, in what historians have called a Cold War proxy. The US had been providing aid and advisors to South Vietnam as early as the 1950s, and by 1968 had more than 500,000 troops in the country.

Each side suffered and inflicted huge losses, with the civilians suffering horribly. During the course of the war, the US lost 58,220 while Vietnamese losses were estimated at over 2 million. The war persisted from 1955 to 1975 with most of the fighting taking place in South Vietnam.

In 1973, US participation in the Vietnam War ended in a cease-fire and a withdrawal that included promises by President Nixon to assist the South in the event of invasion by the North. However, in early 1975, when North Vietnamese forces began a full-scale assault, the US Congress refused to send arms or aid.

Leading right up to the fall of Saigon more than 46 years ago, there was a hastily organized rescue effort dubbed "Operation New Life," which was a massive resettlement program for the thousands of Vietnamese fleeing political oppression, poverty, and continued war. The evacuation was organized by scheduled flights from Saigon to reception camps in the Philippines and Guam, and by transport aboard US Navy ships.

Today, these immigrants from Vietnam and their 300,000 or so children, along with their culture and cuisine, are part of the American mosaic. For them, the only option was to flee or to find American friends still in Saigon who could help them escape.

Clarke's narrative takes the reader to the build-up of the evacuation of not only Americans but also Vietnamese who wanted to board the next available flight from Tan Son Nhut airport.

In this sweeping tale of heroism, scores of Americans, diplomats, business people, soldiers, missionaries, contractors, and spies, all risked their lives to assist their current and former translators, drivers, colleagues, neighbors, friends, and in many instances, strangers to escape.

One of the heroes singled out is diplomat Kenneth Quinn. We met much later in Cambodia when he served as ambassador and I was reporting for *The Washington Times*. At that time, I was not informed of his backstory in Vietnam. During his diplomatic career, Ambassador Quinn served as a rural development advisor in the Mekong Delta, on the National Security Council staff at the White House, at the US Mission to the UN in Vienna, and as Chairman of the US Inter-Agency Task Force on POW/MIAs.

A fluent Vietnamese speaker, Ambassador Quinn acted as interpreter for President Gerald Ford at the White House and personally negotiated the first-ever entry by US personnel into a Vietnamese prison to search for US POW/MIAs.

On March 28, 1975, Quinn was back in Saigon on a presidential mission, headed by General Fred Weyand, a past commander of US forces in Vietnam. The capable and dedicated Foreign Service officer first served in Vietnam in 1968 as a district senior advisor in Sadec Province. Later, he found himself at the epicenter of ensuring as many refugees could be offered safe passage to America. Although he was later assigned to the National Security Council in Washington, his love for Vietnam never wavered and was fortified by his marriage to a Vietnamese and connections with her family and friends.

Clarke writes, "Washington was 12 hours behind Saigon, so it was early on the morning of April 28, 1975 when Quinn (now back in Washington), heard that the Pentagon had stopped the flights and was recommending that Ford halt the evacuation of Vietnamese by fixed-wing planes and concentrate on rescuing Americans."

Quinn raced to the office of White House photographer David Kennerly and explained the situation. Kennerly quickly went to the Oval Office and told President Ford that a reliable source had informed him that "thousands of refugees were stranded at Tan Son Nhut."

He knew that Ford treated Kennerly like a son, and because of his swift actions in communicating with the photographer, thousands of Vietnamese were safely flown out of Saigon.

There are so many other heroes that Clarke identifies, including Richard Armitage, Walter Martindale, Theresa Tull, Marius Burke, Ken Moorefield, James Parker, Al Topping, Brian Ellis, John Madison and other Americans, including US Navy personnel who picked up 20,000 refugees stranded at sea.

The US eventually accepted well over a million refugees who resettled in communities across the country. Even in Quinn's home state of Iowa, the late governor Robert Ray, a Republican, faced down many staunch critics when he called openly for Iowans to welcome nearly 10,000 Vietnamese in 1975.

Clarke's book is timely since Americans today are watching another humanitarian crisis unfold at America's southern border, where the stark evidence of a broken US immigration system now reveals death, detention and the separation of families.

That's not to say that in the 1970s the Vietnam refugee resettlement campaign didn't elicit public opposition. However, the past Trump administration waged a deliberate campaign of human rights violations against asylum seekers in order to broadcast globally that America no longer welcomes refugees.

But now back to my story, since I have digressed again like a true Southerner. The subject of the South China Sea and boat people, whether they are classified as refugees or migrants is relevant for all policy experts and for humanity. Special provisions in customs and law relating to safety protect all of us and rescue at sea.[4] The duty to rescue persons in distress at sea is a fundamental rule of international law. It has been incorporated in international treaties and forms the content of a norm of customary international law.[5]

It is not possible to render all the dreams, fears, hopes, and losses that were experienced for those adrift on the sea of freedom. James Agee, an American writer, wrote in *Let Us Now Praise Famous Men*: "As it is, I'll do what little I can in writing. Only it will be very little. I'm not capable of it; and if I were, you would not go near it at all. For if you did, you would hardly bear to live."[6]

In my many journeys back and forth from the US to Vietnam, and there were many over the past several decades; not only did more Vietnamese refugees or "boat people" stories come my way, but I came to know how the maritime regime, and especially the South China Sea, is not just a geopolitical contested body of water but is also infused with humanitarian values that may not be fully appreciated among policy experts. They are, however, understood among all seafarers.

This recognition of shared humanity is surely found in the published images and heartbreak stories of the thousands of Syrian-refugees crossing the treacherous Mediterranean Sea. Similarly, to the fleeing Hong Kong activists swept up in pro-democracy protests who attempted to outrun Chinese coast guard boats as they sped away from Hong Kong's harbor towards Taiwan. Now in our rear-view mirror, we revert to those grainy black and white photographs of Vietnamese who climbed into leaky and unseaworthy boats when they fled Saigon.

I am reminded of poet activist Warsan Shire's poignant lines from "Home," his poem about migration. "No one puts their children in a boat/unless the water is safer than the land."[7]

After the tragic war came to an end, with growing economic hardship and food shortages, more Vietnamese in fishing junks, makeshift rafts, and freighters sought sanctuary in Southeast Asian countries such as Thailand, Malaysia, the Philippines, Indonesia, and Hong Kong.

There are too many stories of heartbreak about refugees at sea to compress into this single chapter, but I am grateful to Luong Thang Dac, a 50-year-old Australian-Vietnamese citizen, who has been piecing together his family's history and connection to the South China Sea. In different time zones, we resorted to email communication. It helps that he obtained a M.A in creative writing at the University of Technology in Sydney. As a result, he saved meticulous notes, scraps of information, photographs, and recordings of conversations with his father, who was a journalist in Vietnam and died in Australia in 2006.

His father, Luong Ngoc Hai, was born in Haiphong in 1914, and lived through several wars, including World War II, and relocated to Saigon from Hanoi after the division of Vietnam in 1954 and the fall of the French at Dien Bien Phu. That final bloody and climatic battle from the First Indochina War was fought between France's Far East Expeditionary Corps and the Viet Minh communist revolutionaries.

"I believe that my dad named me Thang since it translates as 'victory' and so perhaps it was in commemoration of the French defeat at Dien Bien Phu," claims Thang.

Hai was an educated journalist, who spoke fluent French and several Chinese dialects. Unfortunately, he was arrested and sent to Con Dao prison, situated on an archipelago and located about 150 miles from Saigon. He was imprisoned because he participated in the first coup d'état against President Ngo Dinh Diem in November 1960.

In fact, thousands died at Con Dao over the span of a hundred years. The French summarily executed Vietnamese because many were anti-colonialists or political prisoners. The prison constructed in 1861 by French colonists was designed to jail those deemed dangerous to the colonial government.

Tourism companies market the large island and the remaining 15 smaller islands as an example of sustainable tourism and eco-tourism with a marine park that offers some of best coral reefs and diving in the country.

Many visitors to Con Dao reflect that it represents the worst and best of humanity. At this brutal prison, referred to as the "Devil's Island of Indochina," there's in plain view a cemetery where more than 20,000 graves were dug for prisoners by prisoners. Adjacent to the system of prisons built by the French, situated a large lotus pond frequented by migrating snow-white cranes.

The cranes symbolize longevity, immortality, and good fortune. In Vietnamese mythology, the Sarus crane is the bird sent from heaven to ferry to God those destined for eternal life. With its five-foot wingspan, cardinal red head, slender beak, and slate gray body, it is an extraordinary graceful glided bird seemingly on a heavenly mission.

In Vietnamese legends, it's believed that to count a thousand cranes foretold good fortune. But for several decades following the Vietnam War, no Sarus cranes were seen in Vietnam, since the land, especially in the Mekong Delta, was defoliated, barren, and wasted. This magnificent bird, an "ancient crane of good fortune," had deserted the people and their farms. Despite climate change, rising seawater, and rampant industrialization, good fortune is slowly returning and with it, the once endangered Sarus cranes are protected and seen in the lush wetlands of the delta.

It was at Con Dao's harsh prison that Thang's father was jailed from 1961 to 1963, and he was only released when President Diem was assassinated on November 2, 1963 and three weeks before President John F. Kennedy was killed in Dallas.

But I need to return to the thread of this story. Even as a four-year-old, Thang remembers throughout the morning before the fall of Saigon—the distinctive rhythmic, thumping of the US built Bell UH-1 Iroquois helicopters, or *hueys* as they were called, as they ferried Vietnamese and others from the city to the safety of the US Navy's 7th fleet merchant ships, including the frigate USS Kirk.

Officially, a total of 7,013 *hueys* were in service in Vietnam; 3,305 of them were destroyed, leading to the deaths of over 2,700 pilots, crew members, and passengers.

The *huey* has been made famous by its prominent role in Hollywood films, from Francis Ford Coppola's award-winning classic *Apocalypse Now* (1979) to *Platoon* (1986) and *Full Metal Jacket* (1987) among many others. Indeed, the helicopter has become synonymous with the Vietnam War.[8]

Thang's family along with 3,628 refugees on board escaped Saigon on April 30, 1975 on this vessel. It departed as Saigon (now known as Ho Chi Minh City) as the city fell to the People's Army of Vietnam and the Viet Cong with tanks crashing through the Presidential Palace.

Around 2:30 pm on May 2, 1975, after several days at sea, the Danish cargo vessel, Clara Maersk, received a distress signal from the Vietnamese merchant vessel the Truong Xuan. The ship's captain, Pham Ngoc Luy, sent out this message: "Truong Xuan ship carries more than 3,500 Vietnamese fleeing Saigon form the Communists after Communists' invasion. The engine room is deep in water. The ship will likely sink."[9]

These evacuees were from all walks of life: entertainers, doctors, fishermen, lawyers, farmers, merchants, a monk, two priests, nuns, students, and soldiers. There were births, deaths, and a man overboard, who was subsequently rescued when the captain turned his ship around to search for him.

It is now regarded as the largest ocean rescue of Vietnamese refugees at the time. The Danish captain, Anton Olsen, did not hesitate to take the refugees on board the Clara Maersk and piloted them safely to Hong Kong.

In Captain Luy's memoir, he wrote that he could see the Con Dao mountain ranges as the ship Truong Xuan headed south of Saigon. It's hard to imagine what thoughts passed through Thang's father's mind. It was a boat under Diem that Hai was first sent to Con Dao prison, and it was the South Vietnam Navy that returned him to Saigon after his release following Diem's death.

Every day, desperate people crowd into decrepit ships, and often are placed into perilous situations by unscrupulous people-smugglers.[10] In recent years, we have witnessed an armada of leaky rubber dinghies ferrying Syrians on the narrow ribbon of water between Turkey and Greece.

The humanitarian crisis escalates with more than half of Syria's population of 22 million, now adrift and in search of a new homeland.

The promise and hope of a better life have proved to be a powerful clarion call to flee, despite all the risks involved. The toll in human life has always been high for those seeking escape on the seas to a new life. According to the United Nations High Commission for Refugees (UNHCR), between 200, 000 and 400,000 boat people died at sea. The sustained mass exodus from the region and the massive international response to the crisis thrusted the UN Commission into a leading role in a complex, expensive, and high-profile humanitarian operation.

A smaller exodus of Vietnamese who found their own way by boat to neighboring Southeast Asian countries followed the US-organized evacuation in Saigon. By the end of 1975, some 5,000 Vietnamese had arrived in Thailand, along with 4,000 in Hong Kong, 1,800 in Singapore, and 1,250 in the Philippines," according to a statement from the UNHCR.[11]

There appears no quick and ready answer in international law to the thorny question of who has responsibility for taking in asylum seekers rescued at sea and in providing a place of safety for those who are confirmed in their need for international protection.[12]

For the record, not all ship captains honor their maritime duty to rescue those in distress in the sea, even though this practice exists under both treaty and general international law.[13]

Despite this broad reach of rescue obligations, the protection that maritime rescue affords is not comprehensive. Fortunately for the Vietnamese refugees, Captain Pham Ngoc Luy, followed the historical maritime tradition, and was the last person to leave the battered Truong Xuan.

In his memoir, the captain wrote, "A piece of Viet Nam was drifting away, and washed ashore at Hong Kong. I feel sad knowing it would be unavoidable that this piece of Viet Nam would finally be broken into many smaller pieces to be spread all over the world."[14]

5

Magical Realism, Myths and Science Connect Islanders

The sky is now an azure blue and the wind so tame it barely wrinkles the surface of the quiet harbor. Dr. Chu Manh Trinh, a marine biologist, along with some of his graduate students from Da Nang University, are standing at the clean and well-swept dock, along with neatly stacked recycled trash bags, at Cu Lao Cham, located about 20 kilometers off Vietnam's central coast.

In the surrounding harbor, there's a patina of living corals; jewels of the sea that offer an iridescent combination of blue, pink, golden-yellow, and green coloration. Here the residents know that their corals reveal many stories.

Science reminds us all too often, that corals are the canaries of climate change and indeed they face death from many threats.

The 2,3000 islanders are harmoniously connected to their East Sea. The fishers, some in their small wooden trawlers and many in the traditional round woven basket boats or "thung chai" cast their nets and lines for abalone, sea bass, grouper, lobster, squid, and sea cucumber. Their home of more than 1,500 hectares of natural forest houses a critical ecosystem that includes coral reefs, seaweed, and sea grass beds. It's not surprising that in the 1960s, the island was referred to as "Paradise Island."

For these islanders, their East Sea, as they prefer to identify the South China Sea, is widely regarded as a potential flashpoint in the world. With its intractable sovereignty claims, threats to fishermen, exploitation of fish, coral reef destruction, biodiversity, and ecological imbalances, it was most reassuring to travel to an uncontested and peaceful island in the East Sea where traditional fishermen have adopted an ocean ethic as evidenced in their daily conservation and sustainable practices.

I traveled to this island as an invited guest of Professor Trinh and when the launch arrived in the island's harbor waters, I knew instinctively that this was a remarkable place with its clean sandy beaches, forested hills, a nature reserve, and the sea.

As I glanced up at the two peaks at the western end of the island, I took in the rich natural evergreen forest, that is home to *salanganes* or swallows, with their long, pointed moon-shaped wings and a forked tail. But it is their nests clinging to towering cliffs, that have long proven to be a delicacy for Asians, especially among Hong Kong Chinese. When I was living in Hong Kong, I heard that the nests may cost as much as $4000 per kilogram. The prized nests are meticulously built three times a year by swiftlets from their sticky saliva on cave walls and cliff sides, where they raise their young. How many of us knew that saliva was at such a premium when ordering bird's nest soup? Of course, wealthy Vietnamese during their Lunar New Year or Tet, also present the expensive nest as a practical health gift.

The island also offers an extraordinary array of healthy corals, multiple species of tiger shrimp and mollusk in the clear waters surrounding the islands.

"When I first proposed my idea that they needed to stop fishing near the coral reefs, they thought I was crazy," exclaimed marine biologist Trinh, flashing his high-voltage smile.

For these Vietnamese, the abundance of local marine life is evident from the crabs, seaweeds, mussels, snails and the essential small coral polyps resembling overturned jellyfish that absorb the excess carbon dioxide in the sea and transform it into limestone.

The Cham Islands is a marine protected area (MPA) that was established by the Provincial People's Committee of Quang Nam Province in December 2005. Trinh, a 56-year-old Da Nang University biology professor, is largely responsible for mapping out the agreed upon objectives of protecting natural resources, and cultural and historical values of the Cham archipelago. In 2009, the area was designated a World Biosphere Reserve by UNESCO.

"Yes, it took a sustained educational campaign to convince the population that conservation would provide long-term benefits to their way of life," claims Trinh.

In a walk-about the community, Trinh is happily greeted as the "professor" by fishers mending nets, by women in the common open-air market, and by the military personnel stationed on the island. The energetic marine scientist expresses quiet pride in the way these residents have embraced a sustainable model for fishing that has yielded a new economy from ancillary revenue streams created by dive groups, home stays, local tour guides, fish sauce and even forest tea product processing.

Cham, a 300 square-kilometer archipelago is also referred to as "Me Lao Cham" (Mother of Cham Island). It boasts a 518-meter high mountain with three peaks (Ngoc Long, Tien But and Bat Lao), formerly the home of the ancient king of Cham Pa. The region has both large and small islands, including: Hon La, Hon Kho, Hon Dai, Hon Man, Hon Lon, Hon Loi and Hon En, all scattered across the East Sea. Hon En (Island of Bird's Nest) is an island where a swallow's nest industry is one of the main work activities of the local people, producing hundreds of tons of bird's nests bound for Hoi An annually.

This fishing village is not removed from the international spotlight sharply focused on atolls and islands amidst the roiling political waves of the Spratlys and Paracel islands. In these troubled waters, the island's fishermen find themselves connected with an unwanted predicament: a geopolitical conflict that often prevents them from making their livings effectively.

There's a myriad of challenges here, including China's construction of artificial islands, coral destruction, the placement of a mega size oil rig in disputed waters, the ramming and sinking of traditional wooden fishing boats, false references of historic rights, and imposed fishing moratoriums.

Of course, the situation in the Spratly islands offers many complications. The more than 700 islands and reefs have been claimed by six countries— China, Malaysia, Vietnam, Philippines, Brunei and Taiwan—for more than thirty years and for different reasons. However, what's clear is that the maritime control of the South China Sea has dual dramatic impacts on freedom of navigation and on the pursuit of resources.

Nevertheless, it does not help conservation or science cooperation when Dr. Wu Shicun, President of the National Institute for South China Sea Studies, exclaims that "strict ecological protection measures are guiding China's construction [activities] out on the reefs." Apparently, in a geopolitical misstep and to assuage environmental concerns, Wu added, that the construction was on reefs that "are already dead."

On July 15, 2016 the Permanent Tribunal of Arbitration in the Netherlands strongly condemned China for the serious and permanent environmental damage it has inflicted to coral reefs and to wildlife in the South China Sea. The International judges took aim at China's great "water grab of atolls in the South China Sea"—not least for its wanton destruction of over 100 square kilometers of healthy coral reefs, dredged and ground up to build artificial reefs, coupled with their efforts to catch endangered sea turtles and giant clams since their shells are used for decorative jewelry.

Professor Nguyen Chu Hoi, senior lecturer at Vietnam National University and former Deputy General Director of the Vietnam Sea and Islands Agency, knows that the blue benefits provided by the sea are only possible by maintaining healthy coastal and marine ecosystems. In a South China Sea program held in Hanoi, he shared with me his understanding of why these marine protected areas are essential. "These environments are critical to sustaining social and economic development in the ASEAN region, as well as to protecting globally significant ecological service values and resources for the world."

Environmental policy seems to be a tool that Vietnam's Communist Party is using to shape both a blue economy future and eco-tourism.[1] Just as Ho Chi Minh, the nation's most well-known revolutionary leader, once remarked, "Forests are gold; if we know how to protect and develop them well, they will be very precious." He could have easily transposed the term forests to the East Sea. Perhaps, the crucible of the ongoing degradation of the environment, among all claimants in the contested sea, is bringing about an unexpected interaction between citizens, ideas, and objects, namely fish and their rapid decline.

Vietnam and China have a complex relationship going back more than 2,000 years, including several periods of Chinese imperial occupation that were ended by Vietnamese uprisings. In the troubled South China Sea, hostilities spilled over the Johnson South Reef in 1988, when Chinese soldiers killed over seventy Vietnamese soldiers, and China raised its flag over the barren rock.[2]

In Vietnam, the roughly 160,000 members of the Cham ethnic minority, whose forebears dominated the South China Sea for more than a millennium, are now quietly on the sidelines of the escalating conflict over fishing rights and sovereignty claims.

Today the Cham remain wary of engaging in such disputes, the present territorial wrangling only serves as a reminder of the symbolic and economic importance of their East Sea and of the Cham culture that was once enriched by trade across it.

The Cham's vast trade network extended northeast to China, Taiwan, and Japan and south to Malaysia and Indonesia. Today, the Cham and the Kinh descendants have witnessed China's oceanic plundering, coastal pollution associated with industrialization, and the destruction of more coral reefs.

According to my friend, Dr. Trinh, the residents now on the protected island are no longer Cham but are Kinh people, Vietnam's majority ethnic group. I also learned that there were nearly 5,000 residents living in this Eden-like setting before 1975, mainly from the mainland districts of Hoi An with many fleeing the war and finding boat passage to Cu Lao Cham.

"After the American war more than half of the residents returned home," claims the advocate for marine protected areas. The population never increased because the young people are seeking jobs and opportunities on the mainland.

Vietnam has adopted marine protected areas to address present and future food security issues. Since the 1960s, the number of fish species in the South China Sea has markedly declined from 487 to 238.[3] Vietnam's marine scientists like Professor Trinh, joined by Dr. Nguyen Chu Hoi and Dr. Vo Si Tuan, are in full agreement that stewardship of the ocean is an essential life force for the future of their nation and the region.

Vietnam's S-shaped coastline stretches 3,200 kilometers, excluding islands and encompasses a total coral reef area of 1,270 kilometers. Hanoi's political leadership recognizes what has long been acknowledged that coral reef conditions are under assault and declining. For that matter, the corals of the entire South China Sea cover an area of 30,000 square kilometers and provide a livelihood for hundreds of thousands of fishers.

It's noteworthy that in June 1985, Vietnam adopted a National Conservation Strategy as a key component of an overall plan for national survival. It emphasizes the need to reduce human population growth, and to increase forest cover to ensure improved soil and water conservation. The original protected areas system was designed almost exclusively to protect upland forest, and largely ignored marine, coastal, wetland, and lowland forest environments.

Vietnam now finds itself at a difficult crossroads between economic development that will bring prosperity to its people and conservation initiatives that maintain the rich beauty and natural heritage of the country. What is encouraging is that the local provinces along with Hanoi are listening to the scientists and willing to adopt their conservation and sustainability new initiatives that partner NGOs, local government, and private businesses—all working in concert towards a more sustainable future.

Since 2010, Vietnam has embarked on an ambitious initiative to create "national marine protected areas." As a result, the country currently has eight of these areas established with plans to add eight more in the near future or by 2020. The goal is that this state-led environmentalism will create a transformative mindset among the nation's younger citizens and their relationship to the sea.

Vietnam's rapid development from one of the five poorest countries in the world in 1985 to one of the world's fastest growing economies has had dramatic environmental consequences: rivers dying from industrial pollution, biodiversity loss, and increasing air quality issues in not only Ho Chi Minh City but also Hanoi. As a result, more young social media inspired youth are forming organizations like "Be the Change Agents" to build support for a cleaner environment.

Cu Lao Cham's MPA has over 277 coral species, 270 reef fish species, four lobster species and 97 types of mollusks. While not as significant in marine biodiversity as the Great Barrier Reef or the Coral Triangle in the South China Sea,[4] it's now recognized as an excellent eco-tourism model without any of the typical and associated problems of pollution: increased rate of exploitation, untreated sewage, harvesting of coral for use in construction, the aquarium and shell trades, careless use by tourists and, more generally, increasing population density on the island.[5]

Some marine scientists would quickly add that there are obstacles to creation of MPAs, including rapid economic development, a growing population with low environmental awareness, financial constraints on government budgets, and opposition from people who fear the loss of livelihood, limited socioeconomic data, and at times an unresponsive and poorly trained bureaucracy.

"Protecting ecosystems, sharing natural resources and diversifying livelihoods are key to strengthening the Cham Island communities," Trinh added, as we stopped at the island's museum to examine village artifacts on display.

During my time spent on the island, I was always awakened at 6:00 a.m. While in the homestay of Mrs. Nguyen This Van, to the blaring community-wide loud speakers placed on poles throughout the village, announcing the start of the new day and daily news. Residents receive their news from the Commune leaders. This includes the all-important daily weather forecast and general reinforcement of conservation and sustainability objectives, like "no plastic on the island" and "keep the streets and pathways clean."

This marine protected model offers scientific evidence to encourage even China to rethink its diplomatic strategy and to move away from its previous "development first, treat pollution later" economic path. In fact, many observers believe that the timing is excellent for a shared sense of urgency associated with the unabated marine resource demands and coral reef destruction. Perhaps Vietnam's renewed attention to conservation, sustainability, and environmental rule may succeed in challenging China to embrace a distinctive environmental diplomacy.

The residents never fail to appreciate their culture, history and stories. Some of their colorful stories are as visible on the surface as the parrot fish with their grinding teeth ripping chunks of coral from the reef. Other fishers require some coaxing for their sea tales.

It seems odd that only a few weeks before arriving at Cu Lao Cham, I was in a modern Cineplex theater in Hanoi watching director Ron Howard's *In the Heart of the Sea*- adapted from Nathaniel Philbrick's book of the same title about the 1820 destruction of the whaling ship Essex by one "Moby Dick" size sperm whale. Now here I was, listening to several of the local fishermen regale me with stories about benevolent whales.

Historical whaling and stranding records, the South China Sea is home to more than one-third of whales in the seas, all of which are listed in the Appendix of the Convention on International Trade in Endangered Species of Wild Fauna and Flora (CITES).

Recent data gleaned from the ecological knowledge of local fishers add to the evidence that part or all of the South China Sea is an important cetacean area with high species diversity that deserves special conservation attention but has been previously overlooked. That's specifically why I was speaking with these fishermen about their cetaceans.[6]

Cham beliefs run as deep as the ocean about the mythic properties of whales, which have been protecting these fishermen and their ancestors for many generations. According to scholar Sandra Lantz in her book, *Whale Worship in Vietnam*, upon discovering a dead whale at sea, fishermen will tow it to the harbor and make sure that it receives a proper burial. The ceremony or belief is called "Ca Ong" or 'Mr. Whale worship.[7]

The locals and other Vietnamese fishermen hold these special ceremonies not only for whales, but also for dolphins and porpoises since they all belong to the order of cetaceans. Fishermen never want to kill a cetacean since they believe that bad luck will result from this barbarous act.

Nguyen Toan Hoang, a 60-year-old fisherman, who was born on the island, described his belief. "Our fishing is still good not only because we protect our coral reefs and our whales. They cannot be hurt or our fishing will suffer."

There are nearly 20 different cetaceans found in the South China Sea and on the Vietnamese coastline, including: the Humpback Whale, Pygmy Sperm Whale, Dwarf Sperm Whale, Rough-Toothed Dolphin, Indo-Pacific Humpback Dolphin, Indo-Pacific Bottlenose Dolphin, Pan Tropical Spotted Dolphin, Spinner Dolphin, Striped Dolphin, Common Dolphin, Melon Headed Whale, Pygmy Killer Whale, Short-Finned Pilot Whale, Irrawaddy Dolphin, Finless Porpoise and Dugong.[8]

Unfortunately, the whaling industry still exists and as a result whales are in danger of extinction. This is especially true for the North Atlantic Right Whales, the rarest in the world with only 500 remaining. They grow to lengths of up to 50 feet and can weigh 70 tons.

Despite the 1986 International Whaling Commission (IWC) ban on commercial whaling, some countries refuse to end their whaling operations. Japan is one of the most serious violators. They launched their scientific whaling program after the ban went into effect and it is widely regarded as a cover for its ongoing commercial whaling operation. The Japanese whaling fleets depart twice a year. In the North Pacific, these whalers kill up to 200 Minke whales, 50 Bryde whales, 100 Sei whales and at least 10 Sperm whales under the questionable banner of scientific research.[9]

The numbers are truly staggering. Whales have existed for more than 50 million years. Yet, according to the Natural Resources Defense Council (NRDC) report, "Net Loss: The Killing of Marine Mammals in Foreign Fisheries", scientists estimate that more than 650,000 marine mammals are killed or seriously injured in foreign fisheries after being hooked or entangled or trapped in fishing gear. Some of the harm is intentional—as is the case when fishing fleets use of massive gillnets set upon dolphins as indicators that fish are present- while other harm is incidental, as when Right whales are entangled in crab and lobster pots.

These inhumane actions bring us straight back to Cu Lao Cham, where marine mammals are both protected and worshipped. The Vietnamese have several key terms to describe belief: *Tin*, which translates as "to believe" and *Sur tin nguong*, "a belief in". What is clear is that all the residents on the island express deep respect for the whale. It would seem that in Vietnam's national character, water or rather the East Sea, is sacred and part of their natural consciousness. This is reflected in their long-standing historical maritime tradition. This explains why their ancestors' deep-water life is infused with legends and myths.

I was told that the Vietnamese fishers respect all whale species, but only worship the *Chuong* or the Sperm Whale. If a whale washes ashore and is stranded, the fishing community provide all the help imaginable to attempt to rescue it. If dead, the villagers treat the carcass with reverence, and then use its bones to build temples. Generally, a funeral is organized for the animal, placed in a shroud and buried with respect. Later, the bones will be dug up, carefully cleaned and placed in a temple for worshipping.

For Vietnamese, failing to do so can result in a torn fishing net, a poor catch, a broken engine, a typhoon or even illness.

Lantz's authoritative study reveals that Ca Ong is both a harpoon mark of fear and reverence. Of course, living whales are also revered. Followers of the cult never hunt the large marine animals, which they regard as giant fish. The carcasses of whales that have died naturally are buried with great respect amidst ritual ceremonies. After three to five years, the bones are exhumed, shrouded and carried to the whale temple to be worshipped, as if the creature was a part of their community.[10]

Whale funerals always attract large crowds but make the authorities uneasy. Vietnam's constitution condemns "superstition", but the whale cult is deeply rooted in the country's culture.

This veneration has resulted in the construction of numerous temples, like Hai Tang Tu, a small but beautiful pagoda, built over 250 years ago in a peaceful small valley with rice fields and plenty of vegetables planted nearby. There are dozens of temples in the country, which are dedicated to worshipping whales that have died of natural causes near the coast.

Fishermen speak quietly and respectfully about both their cultural-mythic legends and stories of how the whale or large dolphin has saved their lives while fishing off the coast of their island and sometimes even pushing their boats back to safety during a fast-moving storm. Since whales were never hunted in Vietnam, there are many stories like the one associated with emperor Gia Long (1802–20), who issued several decrees conferring the title "Admiral of Southern Sea" on whales. In recent history, theatrical productions were mounted along the beach for the pleasure of local villagers and for insuring a successful fishing season.

All along Vietnam's coastline, fishing communities worship giant whales, which they view as their guardian angels, an almost religious phenomenon that is unique to the country. However, some policy experts have taken these myths and new historical studies and attempted to politicize them in order to validate Vietnam's sovereignty claims in the Spratlys and Paracel islands.

Scholars like the late Professor Tran Quoc Vuong, has argued that Vietnam's nation-building process since "the dawn of time was connected closely water.

It is easy to find water related history in myths, historical stories, private and oral histories on this period of time."[11]

He wanted to shape some maritime archetype that gave Vietnamese a multi-ethnic identity but that drew upon ancient culture that offers meaning and resonates today among all Vietnamese and that is their love for the East Sea and their fishermen.

But the scholarship drawing any straight line from Vietnam's sea-oriented policies related to medieval marine commerce and trade seems far removed from the lives of the local fishermen.

A few of the older fishermen, seated at a nearby café, close to the local market, and overlooking Cu Lao Cham's pristine harbor, with its colorful corals, told me about how whales spend the majority of their life in complete darkness, but they assured me that these great creatures offered them much knowledge about the currents of ocean life. But just as Herman Melville's novel *Moby Dick* through his character Ishmael, makes clear in the chapter, "Cetology," these islanders also know that any attempt to explain the world in neat confident systems is futile.

Over the centuries, the oceans have given birth to many myths, legends and events still not understood by man. Even the ancient Greek historian, Plutarch wrote: "To the dolphin alone, beyond all others, nature has given what the best philosophers seek; friendship for no advantage. Though it has no need of man; yet is a friend to all men and has often given them great aid."

Since the South China Sea exceeds over 7, 3000 islands with offshore archipelagoes such as the Paracels and Spratlys, where there is a distinctive marine environment consisting of submerged atolls sometimes with emergent islets, but always with lagoons and deep parts of the ocean that connect the scattered coral reefs via ocean currents. And it is these lagoons that hold the nutrients needed for biological production and serve as habitats for marine species.

In his book, *Vietnam Sea and Maritime Trade*, Professor Nguyen Van Kim, who is also the director of the Centre of Sea and Islands Studies at Vietnam National University in Hanoi has generously written that the sea is the starting point, as well as the home of many legendary figures in the memories of Vietnamese. His research includes a study of the 14th century Vietnamese semi-fictional work, Selection of Strange Tales in Linh Nam written by Tran The Phap. In this tale, the sea god Lac Long Quan, helped fishermen and sea merchant boats against a sea monster (Ngu Tinh). This sea monster resembles a giant fish in the East Sea.[12]

According to the tale, when a cargo or fishing boat passes nearby, the monster rocked its tail to cause huge sea waves, causing the boats to capsize. The sea god (Lac Long Quan) is transformed into a merchant boat and traveled to the place where Ngu Tinh frequented. When Ngu attempted to swallow the boat, the sea god Lac Long Quan used a burning iron bar and magic sword to destroy it.[13]

Vietnamese who live along the coast are grateful for the sea in providing them with a source of life. It is their beliefs about the power of the ocean that have led to the emergence of their form of magical realism. For example, a seventeenth-century Cham manuscript from central Vietnam recorded that a local man, returning from Kelantan after a period of studying Malay magic, was drowned when his boat sank in a storm. Apparently in discovering the drifting body, a whale carried the corpse back to shore, and was showered with honors by a grateful community.[14]

The South China Sea's claimant nations need to take a page or two from Cu Lao Cham's history, myths, and science adoption, because they link their valued ocean relationships from ancestral beliefs, myths and sustainable marine ecosystems. These islanders indeed practice what Dr. Sylvia Earle, the legendary marine scientist and the former Chief Scientist of the National Oceanic and Atmospheric Administration (NOAA) believes, "With every drop of water you drink, every breath you take, you are connected to the sea."[15]

6

Vietnam's Rice Bowl Threatened by Climate Change and Dams

Over loaded trucks barrel down the National Highway from Can Tho, Vietnam's fourth-largest city, rumbling past industrial campuses and export processing zones, kicking up more dust from a newly landscaped Chinese paper and pulp mill. Across the road, the delta's dense jungle and mangroves spill over its banks.

For generations, rice farmers and their families harvesting their shining emerald rice paddies have relied on its thousands of river arteries along the once fertile delta; but there is a perfect storm blowing in that is challenging their livelihoods. Nguyen Hien Thien, a 64-year-old rice farmer, shouts, "too much water, and more often, too little." Although, he may as well add, too much pollution to his daily, unheard cry.

The unpredictability of the rains, coupled with an alarming rise in pollution levels, and China's construction of upstream dams is transforming life here.

The delta, formed by the Mekong River, rises on the Tibetan plateau and flows 2,600 miles before dividing into the Cuu Long ("Nine-tailed Dragon") and spills into the South China Sea. Despite the abundance of water that could supply the area, the delta's network of rice paddies, marshes, and canals is dramatically impeded either by too much water in the flood season or too little during the low flow. An agricultural wonder, the Mekong Delta produces half of Vietnam's rice, but now faces growing environmental challenges.

The river's slow-moving current from China, Myanmar, Laos, Thailand, Cambodia, and spilling into Vietnam's Mekong Delta serves as an artery for riparian states, where aquifers, ecosystems, and resources are shared. For centuries, the Mekong River has provided fish, indispensable silt, sediment, and seasonal floods to nourish a world of finely balanced ecosystems. The region is home to 60 million ethnically diverse residents and thousands of species of endemic flora and fauna, including the large Sarus crane, Irrawaddy dolphins, and giant catfish.

The Mekong tributaries have long shaped one of the world's renowned "water civilizations" in the downstream delta characterized by agricultural wonders, floating markets, and attendant distinct lifestyles and beliefs.[1,2]

Upstream dams built by China are a prime culprit, through changing weather, saltwater intrusion, biodiversity depletion, rising sea, and industrial pollution pouring into the delta's rivers are wrecking the ecology of the once fertile delta, historically the rice bowl once considered the fertile rice bowl for over 20 million people in southern Vietnam and a major contributor to the country's vast rice export business, which holds a fifth of the total world export market.

Current evidence reveals that upstream dams are causing irreparable damage to the delta, altering fragile ecosystems and wrecking the livelihoods of the 2.3 million farmers along who farm along the Mekong river and the canals in Vietnam's Mekong Delta. Since 2010, the Mekong Delta has witnessed record repetition of devastating drought every four years.

The 2016 historic drought, followed by saltwater intrusion, cost Vietnam 15 trillion VND (669 million USD) due to the heavy toll on agricultural production.[3] This drought, the worst in a century, caused dire humanitarian impacts, resulting in almost half a million households without access to fresh drinking water and a forced migration of farmers to urban areas in search of jobs.

The delta, a low-level plain less than 10 feet above sea level, is crisscrossed by canals and river systems, where boats, homes, and floating markets co-exist. Some families still recall that South Vietnam's delta proved to be a quagmire for Vietnamese and Americans who fought and died there during the prolonged "American War." It was during this war that American soldiers saw first-hand clusters of thatched roof hamlets surrounded by verdant rice paddies, banana groves and thick vegetation.

Hai Thach, a tired-looking 65-year-old farmer, watches like a sentinel, as the salinity of the water on his land rises—land he has cultivated since he was a boy for rice, coconuts, and mandarins. For Mr. Thach and an increasing number of residents, the search for fresh water often means a half-day upstream to collect enough for drinking, washing, and cooking.

The balance of river and sea is dramatically shifting. Droughts in the delta continue to devastate food supplies, fueling the rancorous debate on China's upstream "run of the river" activities that include six hydropower dams upstream. The dams are not only preventing the flood waters from reaching Vietnam's lower Mekong Delta, but also hold back the flow of sediment that enriches the soil and provides food for fish.

In the 2019–2020 drought season, almost 2 million people inhabiting coastal areas suffered extreme freshwater scarcity.[4,5] Record low water levels in most of waterways and rivers are causing saltwater intrusion that is reaching far inland, up to 75 to 90 kilometers (approximately 46 to 56 miles) from the estuaries, wiping out crops and contaminating water supplies. Because of the prolonged drought coupled with an extreme buildup of salinity from sea-level rise, five provinces in the Mekong Delta declared a state of emergency in March 2020.

Increasing numbers of scientists and environmentalists, including youth environmental groups, see a direct linkage between China's hijacking of the flow of the Mekong River and the interruption of the natural cycle that feeds the ecosystems. Beijing's water diplomacy program is flawed since their dams weaken the river's flow and allows seawater to intrude farther upstream.

Beyond the dams, dramatically shifting weather patterns also pose a threat.

"Climate change will be the most significant environmental impact in the future," said Mekong Delta ecologist Nguyen Huu Thien. "Flood and inundation are increasing frequency; and magnitude, following sea water intrusion with high tide, contaminated soil, sea level rise, seasonal tropical storms (increasing) as a result."

The World Bank, the Vietnamese government, and private enterprises are teaming up on a $400 million plus program to aid nine provinces dealing with extreme weather patterns and the problems posed by the Chinese dams. Vietnamese government planners now project that about 45 percent of the Mekong Delta will be affected by saltwater intrusion by 2030 if hydropower dams and reservoirs continue to stop water from flowing downstream.[6]

With its 2,000 miles of coastline, Vietnam presents a major environmental and food security challenge, especially in the Mekong River Delta where 22 percent of its population lives. Rising seas are inundating low-lying regions, especially in the delta.

While the local governments remain uncertain about how to respond to the repeated devastating droughts and seek solutions to sustain local ecosystems and the livelihoods that depend on them, extreme weather conditions are worsening the Mekong Delta's environment and profoundly changing local distinct cultures and lifestyles. Decreased downstream flow of Mekong water in recent flood seasons means that the nutrients needed to feed the biodiversity and the water required to refill aquifers—which is the main water supply in the dry season—are not reaching critical wetlands. The hydropower dams cause depletion of natural fish resources and imperil fishing villages along the river.[7]

It seems that every year, the Mekong River Commission appeals for an emergency release of water from China's upstream Jinghong Dam in Yunnan Province to meet the delta's desperate needs. While this action was praised, it only serves to underscore the control that Beijing exerts over the Mekong, the 12th largest river in the world, against a backdrop of traditionally suspicious relations between Beijing and Hanoi.

Washington policy makers recognize that the heart of the water diplomacy debate remains squarely on water resource development in Southeast Asia.

At an online Mekong Environment Forum symposium held on April 27, 2020, Brian Eyler, from the Washington-based the Stimson Center's Southeast Asia Program Director and author of *Last Days of the Mighty Mekong*, showed satellite data that demonstrated that China's record of impounding water has led to far more serious droughts for downstream countries. By focusing on a United States government–funded study from Eyes on Earth, a research initiative monitoring water resources based in Ashville, North Carolina, Eyler highlighted that evidence from the physical river gauge from the Mekong River Commission and remote sensing confirm that the ongoing drought is the result of the Chinese water management policy. The data reveals that from 1992 to 2019, satellite measurements of surface wetness in China's Yunnan Province suggests that the region actually had slightly above-average combined rainfall and snowmelt from May to October 2019.[8] "When drought sets in, China effectively controls the flow of the river," claims Eyler.

Can Tho University research scientists say they are deeply concerned about the environmental risks posed by a string of new large dams. Ecologist Mr. Thien observed, "The hydropower dams in the upper Mekong River are sinking and shrinking the delta."

Furthermore, the unique floating nomadic lifestyle formed by the fusion of diverse cultures, ethnicities, food habits, and traditions of thousands of floating traders is disappearing as they travel on their houseboats to all corners of the delta searching for a stable way of life.[9]

Hydropower projects alter natural flow patterns and disrupt fisheries and other ecosystems. The network of dikes built by the government seems to also permanently alter the way nature accommodates the water supply in flood and dry seasons.

A two-year Mekong Delta study commissioned by Vietnam on the impact of Mekong dams was roundly criticized as offering too much technical modeling and failing to connect to the real fears and perspectives of rice farmers and fishers. As a result, some younger researchers joined with local to create a new NGO in the delta.

Nguyen Minh Quang, a 33-year-old Can Tho University lecturer and writer on conflict studies, and I co-founded an independent non-government organization, the Mekong Environment Forum (MEF), to encourage more young people from Can Tho and from ASEAN countries, to immerse themselves into the daily lives of rice farmers and aquaculture communities to better introduce sustainable practices.

Dr. Philip Minderhoud and Dr. Sepehr Eslami Arab from Utrecht University are research members of the Rise and Fall Project, a five-year joint research project by Utrecht University and Can Tho University on land subsidence and salinization in both groundwater and surface water in the Mekong Delta. They presented their research findings at our Can Tho, Vietnam-based MEF online symposium. Their Mekong Delta study confirmed that saltwater intrusion is less than 5 percent due to climate change. Instead, it is principally due to hydropower development.[10]

According to the two researchers, the fluvial sediment supply in the Delta has dropped nearly 90 percent because of the upstream dams. Their studies and others highlight that upstream hydro-infrastructure developments impact basin flow regime biology, bed and bank stability, biodiversity, fish productivity, and sediment and nutrient transport. Depletion of sediment flow is increasing and leading to the fast erosion of riverbeds and banks far beyond climatic trends. When the dams regulate the flow of the Mekong and kill the flood pulse, the Tonle Sap lake can no longer function as a historical flood retention reservoir and thus fails to supply needed water to the Mekong Delta. This "ruin of the river" destructive pattern has long been recognized; however, new data has affirmed that the impacts are as severe as feared.[11]

"Of course, local social media users are now actively participating in the discussion of the pressing delta environmental issues and this includes everything from the upstream dams to pollution associated with the industrialization in the region," says Quang.

During our meeting in Can Tho, Quang, a fearless motorbike driver, boldly told me to jump on the back of his Honda motorbike and we headed straight to the newly constructed $1.2 billion Chinese-owned Lee & Man paper and pulp mill complex, located beside the Hau River, one of the most important waterways in the Mekong Delta.

We were not allowed to enter the plant and were summarily turned away at the high-security gate entrance to the plant. Despite environmental concerns about the plant's wastewater treatment of concentrated toxic chemicals, it is open for business.

Nevertheless, the plant's operations have been under scrutiny by local authorities. The environmental reports generated by Can Tho University environmental scientists have placed the plant on notice to change their factory's wastewater discharge policies. Under pressure, the factory has taken measures to improve their technologies and have employed EU-recognized environmental monitoring and protection equipment.

After the 2016 Formosa steel plant environmental disaster on Vietnam's central coast in Ha Tinh province, resulting in the death of millions of tons of fish from the plant's dumping of untreated wastewater into the sea, authorities were pressured by locals to call for another environmental assessment impact on the newly built paper and pulp mill.

Even the powerful Vietnam Association of Seafood Exporters and Producers (VASEP) urged an assessment on the plant's environmental impact on the vital Hau River.

The total area of the Lee & Man paper and pulp mill stretches more than 247 acres, and over 610 families had to be relocated during its construction. Local water authorities lack knowledge or are untrained to address sensitive environmental issues like effluent wastewater.

In the early stages of its operation in 2016, the Lee & Man paperboard plant expected to use 17 types of chemical for its production, ten of which are said to have "average and high toxicity," said Professor Le Huy Ba, former head of the Institute of Environmental Science, Engineering and Management.

As a result, the Ministry of Natural Resources and Environment requested that the paper and pulp mill halt its operations due to public concerns over its environmental impact. The Vietnam Association of Seafood Exporters and Producers (VASEP) believed that the large volume of toxins could disrupt the Hau River's ecosystem by destroying seafood resources and seriously affect aquaculture in the Mekong Delta, which accounts for 70 percent of the nation's fisheries and aquaculture production.[12]

This decision was based on science. Associate Professor and Dr. Le Trinh, head of the Institute for Environmental Science and Development, said the plant may use sodium hydroxide (NaOH) and chlorine (Cl2) in bleaching pulp, which will result in a chemical compound known as 2,3,7,8-Tetrachlorodibenzo-p-dioxin.

"This is the most toxic chemical compound humans can produce," Trinh underlined.[13]

Rice farmers in these communities recognize that these plants discharge toxic waste and they do not want to become the next cancer village. Some older villagers also reveal that their farming land was revoked to develop the expanded eight industrial zones in the delta. Thus, they worry that their children have no choice but to become workers in these Chinese-backed industries.

It was only a matter of time when the Vietnamese government will no longer be able to trumpet their "rice first" agricultural policy. That day has arrived. I learned from my Mekong Delta science friends at Can Tho University science friends that Vietnam, the world's third biggest exporter of rice, began 2021 by buying the grain from rival India for the first time in decades amid limited domestic supplies and worrisome food security issues associated with the ongoing Covid-19 threats.

7

The Fishing Battleground: Caught between Covid-19 and China

A mid the staggering statistics of more than 4 million Covid-19 deaths, the unprovoked sinking on April 2, 2020 of Captain Tran Hong Tho's wooden trawler by a Chinese coast guard vessel in the disputed South China Sea was lost in the storm of the current mounting pandemic.

While Tho and his crew of eight survived, it is clear there's little time for the world to pay attention to this wreckage against the current background of the global threat of the novel coronavirus and the mounting grim statistics of ravaged lives, shuttered stores, schools, and factories everywhere.

In the early morning with a red sky on the horizon, the 33-year-old Quang Ngai province fisherman had already said his goodbye to family and set sail with his crew to earn their livelihood near the Paracel Islands. It's a scene played out every day as these sentinels of the sea raise and lower their nets.

All fish stories begin the same way, and this one is no different. It begins with fish.

The ocean is vast, covering over 140 million square miles. It represents some 72 percent of the earth's surface. Since early recorded history, oceans have served as the passageway for adventure, commerce, discovery and trade. Like the coronavirus, the ocean has kept people apart and connected them.

Despite encompassing nearly 1.4 million square miles, the South China Sea is a marginal sea, which to cartographers and marine scientists simply means that it is the division of an ocean. Along with archipelagos, islands, and peninsulas, it is the place where an ocean surrenders itself to the inevitably of the earth. It is an ocean interrupted, or at the end.

These fishers, bound to communities that celebrate and link maritime heritage to national history, are caught in the middle of disputes between external belligerent parties, which want to monopolize decreasing maritime resources.

The sea is under all kinds of assault: climate change, acidification, coral reef destruction, industrial pollution, marine plastics, and overfishing, to name but a few. The Spratlys, an archipelago, comprised of hundreds of reefs, sandbars, and tiny atolls, sprawl some 160,000 square miles. These islands are surrounded by rich fishing grounds and potentially by gas and oil deposits with one of the busiest international sea lanes nearby. Vietnam and China have for years been embroiled in a dispute over the stretch of water, referred to as the East Sea by the Vietnamese.

For centuries, the South China Sea has provided abundant fisheries for its coastal countries. Flushed all year round by many large rivers, the flat and shallow sea beds of this body of water are among the world's most productive fishing grounds.

As a result, there's increasing competition for the more than 3,000 species of fish swimming through these waters. But overfishing, along with illegal and unreported catches, has already collapsed coastal fish stocks, forcing trawlers to go further from home.

But for fishermen, whether they are from Ly Son Island or Tanmen, China poses only increased threat and fear because of the coronavirus. There's much speculation about the origins of the virus but it's certain now that the initial outbreak was in Wuhan, China. Additionally, the onset of the virus has caused closures of so many seafood markets and others that have remained open in the region have to limit vendors selling at the market. Ongoing curfews are also limiting the ability of some fishers to travel to and sell at major market centers.

In the complicated endgame blame, most agree that Beijing mishandled the epicenter of the disease by silencing doctors, concealing test results, and granting selective access to the World Health Organization. I read with much interest Lawrence Wright's impressive reporting in the New Yorker, titled, "The Plague Year." His analysis includes irrefutable evidence that China's head of Disease Control, George Fu Gao, lied to the Robert Redfield, the director of the Center for Disease Control and Prevention, when he pressed Gao about whether there was evidence of human-to-human transmission.

"When Redfield learned that among twenty-seven reported cases in [China], there were several family clusters, he observed that it was unlikely that each person had been infected, simultaneously, by a caged civet cat or a racoon dog. He offered to send a CDC team to Wuhan to investigate, but Gao said that he wasn't authorized to accept such assistance."[1] Of course, the US is not blameless and President Trump failures have been splashed over the pages of newspapers. Wright rightfully places missteps from the Centers for Disease Control and Prevention and also the Food and Drug Administration in their failures to roll out tests. In this case, too many technocrats failed miserably in their jobs.

But my thoughts take me back to a lack of transparency that is a fundamental part of the Chinese endgame. This lack of transparency spills over into the South China Sea since China does not provide data on the health of the coral and fish populations at the reclaimed reefs where they have constructed various military-type installations. In a Council of Foreign Relations briefing, Ely Ratner, the executive vice president and director of Studies at the Center for a New American Security, outlines this break or lack of transparency.

"The dearth of public information about China's activities in the South China Sea has hampered regional coordination and abetted China's ability to take incremental steps to consolidate control. Countries in the region currently lack sufficient surveillance capabilities to monitor even their own exclusive economic zones, much less keep track of China's deployments and infrastructure development on its artificial islands."[2]

Meanwhile, apart from the IUU fishing and the maritime incidents involving the fishermen, the present collision course over lingering sovereignty and maritime boundaries becomes ever more entangled. That is not to say that Vietnam does not use fishermen as sentinels or part of a paramilitary force, especially in the Paracels.

Greg Poling, director of the Asia Maritime Transparency Initiative (AMTI) at the Center for Strategic and International Studies (CSIS), claims that over 50 percent of the fishing vessels in the world operate in the South China Sea. Vietnam has over 30,000 fishing boats and China has 100,000 or more steel hulled trawlers plundering the declining fish stocks.

Again, Vietnamese fishermen are not blameless since they too have been frequently caught illegally fishing in Indonesia's waters close to Natuna Islands. The islands are situated on the outermost northern border islands in the

Riau Archipelago, on the west side of the South China Sea. Strategically located between Singapore, Malaysian Peninsula and East Malaysia on Borneo. The sea between these two island clusters offers fishermen exceptional trawling grounds. Naturally, the Indonesian government regards the Natunas issue not as one of sovereignty but of illegal fishing in its EEZ. The incursions by Chinese and Vietnamese fishing boats only serve to dramatize the lengths that fishermen travel to find their fish. The South China Sea remains seriously overexploited with fish stocks declining anywhere between 70 to 95 percent over the past four decades. As long as the conflict in the disputed sea continues, it appears nearly impossible to regulate fishing.

China, with its 15 million fishermen, represents nearly 25 percent of the world's total, while ASEAN countries have at least 30 million fishermen, according to the Food and Agricultural Organization of the United Nations (FAO). These fishing figures, the decline of inshore fish stocks, massive government subsidies, and rising consumption of fisheries products at home and abroad only add exponential pressure to the fragile marine ecology, and furthermore increases the potential for more clashes among fishing fleets in the South China Sea.

A decade ago, Daniel Pauly, a professor at the Fisheries Centre of the University of British Columbia, wrote an essay, "Aquacalypse Now" in *The New Republic*. "It's essential that we [act] as quickly as possible because the consequences of an end to fish are frightful. To some Western nations, an end to fish might simply seem like a culinary catastrophe, but for 400 million people in developing nations, particularly in poor African and South Asian countries, fish is the main source of animal protein."[3] Pauly and his colleagues at UBC's Sea Around Us project have devoted resources to identifying and quantifying global fisheries trends.

The recent attack on a Vietnamese fishing trawler marks the second time in less than a year that a vessel has been sunk by a Chinese ship near the PRC-controlled Paracel Islands. As Vietnamese fishermen struggle daily to protect their livelihoods, their fishing grounds, and their country's territory, Chinese paramilitary vessels steal their equipment, damage or sink their boats, abduct them, extort money and beat them.

Chinese have even killed several Vietnamese fishermen. Two decades ago, Chinese boats rained bullets on two defenseless fishing boats less than 20 nautical miles from Vietnam's Central coastline, killing nine and gravely wounding seven others. The attackers then held the survivors and the two boats until the fishermen's families paid substantial sums of money.

Some Washington analysts do suggest the Covid-19 pandemic gives cover to not only China, but also other claimants to leverage the health crisis as a strategic opportunity to sail into disputed waters. John Ruwitch, a respected international journalist, produced an investigative report that outlines that China's fishing fleet is connected to maritime enforcement authorities and has over sixteen satellites in orbit above the Asia Pacific, with more planned. In fact, over 50,000 fishing Chinese fishing trawlers and fishing boats are networked through their Beidou satellite system. Their updates and last satellite launched in June 2020 make it competitive with the US Global Positioning System network. Beidou is named after the Chinese word for the Big Dipper constellation and it took almost two decades to complete.

Additionally, China's financial and technological support of fishing through subsidies contributes greatly, to Beijing's ability to annually assert its maritime claims through their fishing fleets. Also, Professor Ussif Rashid Sumaila, a University of British Columbia Fisheries Centre expert, offers ongoing research studies that China's fishing subsidies continue to aid overfishing. In short, the data claims that Asia leads the world in fisheries subsidies with both China and Japan at the apex.[4]

It's noteworthy that in 2020 China launched over 34 satellites despite the Covid-19 pandemic. In fact, China now has more than 363 satellites in orbit insuring their coverage of the Asia Pacific region. Most experts agree that their Beidou system is a rival to the dominant US GPS and Russia's own equivalent of a Global Navigation Satellite System (GLONASS).

The Beidou program system comes equipped with an emergency button that sends a message directly to Chinese maritime authorities such as the coast guard. These messages offer accurate and detailed data on the location of the identified fishing boat.

The English language website of the People's Liberation Army highlights Beijing's large-scale naval exercises, chronicles not only the sinking of the Vietnamese fishing boats, but also reveals the latest ramping up of military industries in Wuhan.

Despite warnings from the US State Department to focus on combating the pandemic and "stop exploiting the distraction or vulnerability of other states to expand its unlawful claims," the Chinese *Haiyang Dizhi 8*, a survey ship, flanked by Chinese coast guard vessels and at the center of a stand-off with Vietnam last year, has returned to the same disputed waters.

Beijing's state press commentators have suggested that Covid-19 has lowered the response level of the US Navy deployment capability in the

Indo-Pacific. Dr. Wu Shicun, president of the Hainan-based National Institute for South China Sea Studies, wrote in a *South China Morning Post* commentary: "Covid-19 has dealt a body blow to the United States, including its military combat capabilities and deployment. It is reported that the virus has been found in at least 150 US military bases and on four aircraft carriers."[5]

While China's propaganda may help bolster a nationalism wave at home and offer the appearance of exploiting the pandemic in their naval and fishing strategies, the US is highly unlikely to change course in its Freedom of Navigation operations under the Biden administration. The purpose of these naval exercises is to challenge maritime claims, promote a rules-based order, demonstrate support for regional allies and ensure open waters. They also indirectly serve as part of an established US deterrence strategy against China.

If anything during this public health storm, US allies like Australia, Japan and the United Kingdom will stand by the US and perhaps increase their military presence in the region.

While neither China nor Vietnam are short of water diplomacy tools in the South China Sea propaganda war, Vietnam takes pride in their low number of confirmed Covid-19 cases with only 272 reported and no deaths. In Hanoi, the Vietnamese Communist Party has effectively deployed the state apparatus to mobilize security forces and healthcare workers to efficiently quarantine and trace tens of thousands of people.

An outbreak aboard a ship can prove fatal and represents a threat to public health, particularly with fishing trawlers sailing across transboundary waters. With over fifty million people employed in marine fisheries and especially in developing countries, all aspects of fish supply chains have been broken by this pandemic, with rising numbers of jobs, incomes, and food security at risk.

For fishermen, there's no chance for social distancing, only constant vigilance as they gaze out across the vast expanse of sea and sky to the farthest grey horizon.

8

Science Weathers Geopolitics
in the Marginal Sea

In the midst of a global pandemic, US voters chose Joe Biden over Donald Trump and his disruptive politics, paving the way toward a new era of science, reason, and international Covid-19 cooperation. Across the Pacific's blue waters, China, the Philippines, Taiwan, and Vietnam, all rivals in the contested South China Sea, are charting a new direction in maritime cooperation and networked multilateralism.

Despite the pandemic, in the fall 2020, China hosted a two-day international ocean governance program with a roster of speakers from the US and claimant nations, including the Philippines and Vietnam. The theme of the inaugural symposium, "Maritime Cooperation and Ocean Governance", preceded Hanoi's program, "Maintaining Peace & Cooperation Through Time of Turbulence, held in November.

What does all of this signal?

In this sea of opportunities, uncertainties, threats, environmental degradation, it is time for marine scientists to pull together to respond to acidification, loss of biodiversity, climate change, destruction of coral reefs and fishery collapses.

Because of these issues, there's a rising chorus among claimant nations, including China who choose to view the South China Sea as an ideal platform for promoting regional cooperation and for participating in marine surveys. The tide is lifting science research survey vessels above the din of politics and sovereignty claims. It brings to the churning sea more central solutions for the region's long-term peace and sustainability. The respective programs also happen to correspond with the start of the 2021 UN Decade of Ocean Science for Sustainable Development that offers an inclusive, participatory, and global process.

Just as the current pandemic necessitates a harmonized and collaborative approach to clinical testing, scale-up, and distribution, the South China Sea requires attentive science cooperation in fisheries management, ecological conservation of coral reefs, and open access to ocean data. Most reporting and academic papers, and there's no shortage of them, focuses on the conflicts and tangled geopolitical disputes in the region. For the most part, the South China Sea has never been heralded as an epicenter for collaborative science.

Fu Ying, Chairwoman of the Center for International Strategy and Security at Tsinghua University, reminded over 500 on site and online attendees of the "urgency to enhance global governance and improve the efficiency of international coordination mechanisms, which is also the case for ocean governance."

The two-day conference was held in Haikou, the capital city of Hainan Province, and co-organized by the National Institute for South China Sea Studies (NISCSS) and the China-Southeast Asia Research Center on the South China Sea (CSARC). Nearby, the Tropical Ocean Research Institute monitors the growth of corals every day. China now has created the largest coral research and cultivation base.

The irony is not lost on Washington nor Hanoi that the fast-tracked maritime cooperation event was held nearby a new port facility where China's maritime militias are based and is close to the contested Spratly Islands. Beijing has not hesitated to use its array of maritime vessels setting out from Hainan to harass smaller rival claimants, especially Vietnamese and Filipino fishing boats.

For more than a decade, China has vigorously staked a claim to 90 percent of the SCS as its sovereign territory, within the so-called nine-dash line.[1]

The line runs as far as 1,080 nautical miles from the Chinese mainland to within 100 nautical miles of the Philippines, Malaysia, and Vietnam. In 2009, the Chinese government circulated a nine-dash map through a set of notes verbales to the United Nations, borrowing their inspiration from an eleven-dash line map published by the Nationalist Government of the Republic of China in 1947. Bill Hayton's seminal book, *The South China Sea: The Struggle for Power in Asia*, provides a sweeping history of the map sagas. He writes, "No official explanation of the meaning of the line was provided, although one of its cartographers, Wang Xiguang, is reported to have said that the dashes simply indicated the median line between China's territory- in other words, each claimed island—and that of its neighbors."[2]

CSARC focus includes traditional and non-traditional security, marine environmental protection, marine scientific research, safety of navigation and communication at sea, joint development and management of natural resources, and crisis prevention and management. While it's encouraging that the Chinese research center does examine marine science research, its program's primary aim is to serve the national economic and military interests.

Beijing's hard-power maritime expansionism is evident in its modern navy, coast guard, and paramilitary fleet that have continued to ram into fishing boats, harass oil exploration surveys, build military outposts on reclaimed atolls, and hold combat drills.

Despite the deeply entrenched cross-strait issues between the People's Republic of China (PRC) and the Republic of China (ROC), commonly known as Taiwan, Taiwanese research fellow Dr. Yann-huei Song, who participated in the China program, echoes the views of the conference organizers in their stated collaboration goals for a shared and sustainable future. "Accordingly, it is my view that the program reveals an authentic Chinese response to science cooperation by adopting a multilateral approach," wrote Song in an email to me.[3]

In an effort to better understand if China's sustainability overtures is not a concerted soft policy propaganda effort, it's noteworthy to understand how China is framing this sustainability concern and development. As recently as 2015, China promoted maritime scientific research and international cooperation to advance marine development. I reached to a spokesperson for the State Oceanic Administration (SOA) that hosted the annual meeting of the North Pacific Marine Science Organization (PICES) held in Qingdao, a coastal city in east China's Shandong Province. "Global research on ocean science and technology has entered a new stage featuring cross-disciplinary studies and international collaboration," claims Chen Lianzeng, deputy director of SOA. implying that all countries should work together to protect marine environments and to combat climate change.

Beijing knows that with its enormous population and heavy reliance on coal, China is by far the world's biggest polluter, responsible for more emissions than the US and EU combined.

One of the key drivers behind Chinese emissions is the intense urbanization that has occurred as millions of people across the country are leaving rural areas to rapidly expanding cities. Of course, with an expanding middle class population of over 400 million, there's a demand for fish. As a result, China has taken an aggressive approach to the expansion of its fishing industry. Their armanda of 2,600 deep-sea fishing trawlers outnumbers all others. China's overfishing and illegal, unreported and unregulated (IUU) fishing practices are major contributors to the depletion of the ocean's resources. Experts now agree that 90 percent of the fisheries are fully exploited or facing collapse.

Because China has been facing many serious environmental issues, from air, water and soil pollution, the Chinese government produced their own "ecological civilization" (*shengtai wenming* 生态文明) proposal embedded into the Chinese Communist Party (CCP) constitution in 2012. It was endorsed by President Hu Jintao in 2007 before it was cast into stone as a central tenet of the green rhetoric of the CCP.[4]

Among sinologists, the relationship between science and political power has always been a major source of interrogation and debate. Why did the CCP choose the term "ecological civilization" instead of sustainable development? The term "civilization" has a specific political meaning in the rhetoric of the Party since Deng Xiaoping's campaign in the 1980s to promote a "spiritual civilization" (*jingshen wemming* 精神文明), complementing the "material civilization" (*wuzhi wemming* 物质文明) brought about by economic reforms.[5]

This ecological civilization concept frames China's vision of ecological, social and economic development to their traditions. Thus, Beijing's efforts to engage claimant nations in dialogues about the shared sea must be seen, at least partly, as a legitimate gesture, since China has an urgent need to become a modern green economy and to be perceived as a global leader in the 2030 Agenda for Sustainable Development and the Paris Agreement on Climate Change.

Even on the Vietnamese front, Dr. Nguyen Hung Son, who was a participant in the 2020 China maritime cooperation symposium, believes that Beijing's actions point to new directions for South China Sea inquiry and cooperation.

As a panelist in the Vietnam program held on November 2020, I also offered solutions for the promotion of science cooperation during my presentation "Marine Science: How science and technology may impact the good order of the sea". The development of sciences and related ocean technologies helps to expand new frontiers on how citizens understand, exploit, and manage the ocean, suggesting new opportunities for cooperation and new alternatives to address maritime issues and reduce tensions.

However, the shape of ocean science is also influenced by larger geopolitical disputes and economic pressure attributed to competition over oil and gas resources, and commercial fishing. There's surprising reassurance from one of China's leading scientists, Dr. Zou Xinqing of the School of Geographic and Oceanographic Sciences at Nanjing University, who believes that "science collaboration is very important for this region."

These science-focused South China Sea programs and webinars offer a marine stewardship initiative for scientists, communities, and the general public to address climate change, coral reef destruction, biodiversity loss, pollution, and fisheries depletion. The corals are especially stressed by the increase in surface seawater temperature, the decimation of crown-of-thorns starfish, acidification, and pollution of oceans. Marine scientists and oceanographers have recognized for some time that an increasing number of coastal coral reefs have already been degraded by centuries of overfishing and pollution. According to an article in the *Annual Review of Marine Science*, "The challenges now are to identify and maintain the ecosystem functions that are crucial for sustaining coral reefs."[6]

Despite some marine science cooperation between China and Southeast Asian countries, the scale remains limited and there remain a mix of skeptics and champions for purposeful marine science collaborative practices.

"I am skeptical that such cooperation will significantly expand in the coming years since the disputes in the South China Sea serve as a huge obstacle for such cooperation," adds Dr. Li Mingjiang, as associate professor at S. Rajaratnam School of International Studies (RSIS) at Nanyang Technological University in Singapore.

China's noted paleoceanographer, Wang Pinxian explains, "the South China Sea is a fantastic natural laboratory for deep-sea research."[7]

There are competing visions for regional marine science cooperation surrounding the contested South China Sea. The expansion of cabled ocean observatories brings data ashore through the Internet. An open access of information-sharing in the disputed SCS can benefit all claimant nations, especially Vietnam and the Philippines, in the form of ocean governance and fisheries sustainability. More Chinese marine scientists have also been engaged in establishing marine protected areas in preventing marine development from destroying fragile ecosystems and their habitats.

For example, Dr. Yu Weidong, at Sun Yat-Sen University's School of Atmospheric Sciences in Guangzhou readily endorses ocean science cooperation as the best way to move forward in the collection and in the stimulation of common interests in the region to address the challenges of ocean science cooperation. "I have some experience working with partners in the region, mainly through Intergovernmental Oceanographic Commission for the Western Pacific (IOC-WESTPAC)." Along with others, he knows that the challenges of climate change, extreme weather, climatic disaster, and disruptions in marine ecosystems require collaboration.

In email communication, he talks more about his expertise and science cooperation efforts. "My study in the region is around the collaborative

WESTPAC project, Monsoon Onset Monitoring and its Social and Ecosystem Impacts (MOMSEI)." Climate change is increasing the dynamics of Asian monsoon seasons and the impact of rainfall variability.

His study included science cooperation participants from Indonesia, Thailand, Malaysia, and partners from Vietnam, the Philippines, and Myanmar.

While no one denies that China has been responsible for wrecking reefs during their reclamations, over 270 marine protected areas have been established in China's coastal areas at the end of 2017, with a total area of 120,000 square kilometers, 4.1 percent of the total sea area. There are also 35 national marine sanctuaries with a total area of 17,751 square kilometers and 67 special marine reserves with a total area of 7,250 square kilometers, according to correspondence from Li Yunzhou, Ben Yiping, and Chen Yong.[8]

Also, since 2005, the National Oceanic and Atmospheric Administration's (NOAA) National Marine Sanctuary Program has partnered with International Union for Conservation (IUCN), the Vietnam Ministry of Fisheries, Conservation International, World Wildlife Fund (WWF), Danish International Development Agency (Danida), and the US Embassy to provide management capacity training in support of developing a marine protected area (MPA) network in the South China Sea.

Ecologists, environmentalists, and marine protected area practitioners from China, Vietnam, and Cambodia gathered in Nha Trang, Vietnam in 2005 for the first MPA Management Capacity Building pilot program. The South China Sea region was selected from nine regional candidate sites for the very first pilot project due largely to the existing partnership between NOAA, IUCN, and the Ministry of Fisheries.

Of course, scientific collaboration does face hurdles in the disputed South China Sea. One major lesson learned from previous marine environmental protection projects was the need to separate scientific and technical issues from political decision-making nationalist rhetoric and sovereignty claims.

I have spoken often with Professor John McManus, ecologist at the Rosentiel School of Marine and Atmospheric Science at University of Miami, at many South China Sea programs. He has repeatedly presented science based-evidence on the need for science collaboration since so many of the coral reefs have not been studied. He also presented a timely paper, "Offshore coral reef damage, overfishing and paths to peace in the South China Sea," at a Haiphong, Vietnam program in 2016.

"37 percent of the 830,000 reef species globally would lead us to expect that the southern reefs of the South China Sea to hold more than 307,000 multicellular coral reef species. Given the sparse nature of studies in the area, coupled with the isolation of some portions, it would be reasonable to expect that the majority of these have yet to be identified, and a large number of them will be new to science. This indicates that the potential for new medical drugs from the sea from these offshore reefs is particularly high", claims McManus.

Coral reefs are wondrous underwater communities teeming with life and possibilities. Coral plants are important sources of new medicines now being developed to treat cancer, arthritis, human bacterial infections, Alzheimer's, heart disease, viruses and other diseases. The antiviral drugs Ara-A and the anti-cancer agent Ara-C, were developed from extracts of sponges and are among the earliest modern medicines obtained from coral reefs. In China, Japan, and Taiwan, tonics and medicines derived from seahorse extracts are used to treat a wide range of ailments, including sexual disorders, respiratory and circulatory problems, kidney and liver diseases, throat infections, skin ailments, and pain.

The prospect of finding a new drug in the sea, especially among coral reef species, may be 300 to 400 times more likely than isolating one from a terrestrial ecosystem.[9]

McManus, the leading advocate for marine protected parks, ardently believes that reefs of less inhabited island clusters, such as the Spratly and Paracel islands, have previously been under less threat but are now subject to more destructive pressures. One of the most critical issues facing the world is the warming of our planet. Anthropogenic activities have harmed the environment in various ways, and coral reefs have been especially hurt. In spite of the fact that coral reefs occupy only 0.2 percent of the ocean area, they are estimated to harbor about one-third of all described marine species.

Elizabeth Kolbert, winner of the Pulitzer Prize for *The Sixth Extinction*, writes in her newest book, *Under A White Sky The Nature of the Future*, "In the age of man, there is nowhere to go, and this includes the deepest trenches of the oceans and the middle of the Antarctic ice sheet, that does not already bear out our Friday-like footprints." She lays out the facts just as my marine scientist friends do. We should indeed be worried.

According to McManus, surveys taken of the reef flat and lagoon at Thitu Reef and situated in the east Spratly Islands with neighboring North Danger Reef to the north and Subi Reef to the west, reveal that these reefs have been overfished (the piscivorous fish are generally absent), that there was good coral cover behind the breakers and on the outer reef flat: mostly small dense, relatively fast-growing Acropora and Montipora colonies interspersed with some low microatolls of various more storm-resistant and slow-growing species.[10]

The Acropora is the most diverse reef building coral in the world and their high growth rates have allowed these reefs to keep up with changes in sea level. Also, their branch-like marbled morphology in layers of green, light purple and yellow, provides a vital home for other reef organisms, such as fishes, turtles, crustaceans and mollusks.

This focus on the protection of coral reefs matters since a new study suggests that by 2050, most coral reefs around the world will be at risk of experiencing constant depletion of their building blocks—calcium carbonate sediments. More than a few reefs are already showing signs of net sediment dissolution, which dramatically impedes the growth of coral reefs and subsequently impacts the health and biodiversity of reef habitats.[11]

The exploration of data sets from the 1980s reveals that the percentage cover of reef-building corals may well have decreased by as much as 50 percent across the Indo-Pacific region.[12] If there's a failure for science collaboration in the South China Sea, the resulting loss of coral reefs will pose even greater threats to food security and livelihoods.

Perhaps that's why Hainan marine scientists are looking toward the future as they build a coral seed bank to serve not only China's ecological interests, but also addressing that of global coral reef ecosystems.

Professor John McManus, an American ecologist and marine biologist, who advocates for marine protected parks in the Spratly Islands.

A Vietnamese boat builder in Da Nang, Vietnam.

Captain Dang Van Nhan, whose fishing trawler was attacked and sunk by a Chinese vessel.

The author inspecting a damaged Vietnamese fishing trawler in Da Nang, Vietnam.

Credit: "Courtesy of A.P. Moller-Maersk Archives" the rescue of 3,500 Vietnamese refugees from the South China Sea post-Vietnam War.

Dr. Chu Manh Trinh, a marine biologist from Da Nang University instructing students about coastal erosion along the East Sea.

Part II
Ecological Politics

9

Follow the Fish and the Law in South China Sea

Frigate mackerel, red soldier fish, yellow fin, and bigeye tuna are jumping again in the South China Sea after Beijing's three-and-a-half-month summer fishing ban in the roiling seas.

This signals a harvest for the thousands of fishing vessels sailing from Hainan, Haiphong, Da Nang, and Ly Son. Even the US has cast its net into the contested waters through a memorandum of understanding (MoU) between the State Department's Bureau of International Narcotics and Law Enforcement Affairs (INL) and the Vietnamese Agriculture Ministry's Directorate of Fisheries (DFISH).

The MoU strengthens Vietnam's fisheries management and law enforcement capabilities, but policy observers suggest that it is Washington's way to help Hanoi curb China's unlawful maritime expansion. During the COVID-19 pandemic, China established de facto control over the South China Sea by sinking a Vietnamese fishing boat, conducting military exercises near the Paracel Islands, and declaring greater restrictions over freedom of navigation.

This MoU initiated during the Trump administration was seemingly a shot across the bow at Beijing since it appeared to boost Vietnam's maritime capabilities. Key components of the agreement include direct support against illegal "intimidation" of Vietnamese fishermen at sea and fostering greater cooperation to ensure sustainable living marine resources and combat illegal, unreported and unregulated (IUU) fishing. The year 2021 marks the start of what the United Nations hopes will be a pivotal decade for the global ocean. The UN is mounting a massive operation to try to raise awareness of the many problems it faces and to harness the scientific research needed to solve them. It's called the UN Decade of Ocean Science for Sustainable Development. The project's motto is "The science we need for the ocean we want". By 2030 the UN expects the world to have more of both. But the question is what will this campaign achieve and how?

"The world will not allow Beijing to treat the South China Sea as its maritime empire," claimed former Secretary of State Mike Pompeo. He added, "America stands with our Southeast Asian allies and partners in protecting their sovereign rights to offshore resources, consistent with their rights and obligations under international law."

This declaration represented the first time that Washington has publicly endorsed the 2016 decision of the Permanent Court of Arbitration in Hague on the status of certain features in the Spratly Islands and represents a reversal from the previously neutral American policy position on the disputed South China Sea. That favorable ruling for the Philippines challenged China's nine-dash line covering most of the South China Sea. Beijing had never clarified whether the line represents a claim to the islands within the line and their adjacent waters.[1]

The Philippines initiated arbitration in January 2013 under the dispute settlement procedures of Annex VII to the 1982 United Nations Convention on the Law of the Sea (UNCLOS).[2]

In March 2020 in the midst of the pandemic, Vietnam offered a note verbale, which explains Hanoi's positions in legal terms which are compatible with the key finding of the 2016 South China Sea arbitration award. The note objects to China's historic rights in the South China Sea and any other maritime claims that exceed the limits provided in the UNCLOS. It also opposes the "four shas" doctrine, a Chinese endeavor to claim maritime zones from four groups of islands in the contested South China Sea as if they were each a single entity. But most interestingly, the note seems to indicate, for the first time, an official Vietnamese position on the legal status of all high-tide features in both the Spratly and Paracel Islands.[3]

This means that all rights to islands, rocks, and low-tide elevations in the South China Sea are based either on historical claims under international law or on maritime claims under the UNCLOS. China has been making the claim that its history in the SCS spans over two thousand years. However, there are other littoral claimants such as Taiwan, the Philippines, Vietnam, and Japan, with the United States implicated in Japan's claim for historical and other reasons.[4]

Meanwhile, Vietnam's alignment with the US is geopolitically timely since Hanoi is still flagged with a yellow card by the European Commission over IUU fishing issues. In the fall of 2020, Vietnam received a notice of comments from the EU, with an additional six months for review to remove the seafood export penalty. Of course, seafood traceability remains a huge problem that is difficult to identify. Fishery enterprises struggle against strict regulations to prove that imports or purchases at sea must have clear origins. The EU inspection team does confirm that Vietnam has made progress and is on the right trajectory.

Washington's objective is to ensure that Vietnamese fishermen adhere to international maritime rules and not be fearful of outside intimidation from vessels of neighboring states.

While Vietnam has made progress with its transparency on fishery products, traceability issues and overall issues of monitoring fish catches remain. Many Vietnamese fishing vessels continue to fish illegally in foreign waters.

The pandemic distractions enable fishers to fish where there's lax enforcement and backslide into illegal operations. Since 2014, Indonesia has blown up and sunk several hundred fishing vessels, including fishing boats from Vietnam for allegedly violating its waters.

A lot is at stake in fishery revenues. According to Seafood Source, Vietnam had almost $10.5 billion in exports from seafood products in 2019, up by 23 percent compared to 2017 figures. Vietnam Association of Seafood Exporters and Producers (VASEP) believes that the country also has one of the fastest growing fishing fleets in the world, increasing from 40,000 in 1990 to nearly 108,500 in 2018.[5]

If Vietnam was to deliver a credible fisheries management regime, it would return fishing efforts to sustainable levels and eradicate the bulk of illegal fishing by its fleet. It must take steps to target the illegal Vietnamese operators in both domestic and international waters. This includes compliance with vessel monitoring systems that specify the location and movement of fishing vessels.

The recent MoU between Washington and Hanoi does bolster the enforcement of strict limits to fishing efforts and help deter violations. However, for local fishermen, their coastal waters remain overfished and their only alternative is to head for deep waters. It's still too soon in the Biden administration to know if they will increase sail-bys and close-in air reconnaissance in the South China Sea. In 2020, Washington conducted 10 freedom of navigation operations (FONOPs) in the contested waters. FONOPs are designed to challenge

China's claims to maritime rights over several island chains in the region. US warships travel within 12 nautical miles of features claimed by Beijing and always succeed in elevating Chinese rhetoric and the blood pressure of Ministry of Foreign Affairs officials. From Washington's perspective, they reinforce the Department of Defense's policy of ensuring that the American military can fly, sail, and operate wherever international law allows.

"There are too many fishing boats. We must go further and further, [despite] knowing that it's dangerous," claims Dang Van Nhan, a third-generation boat Da Nang captain, who has been casting his long-line nets into the turbulent South China Sea for two decades. Since the fishermen are always the first to address the limits of the sea, their voices are increasingly amplified.

Recent data from the Fujian Ocean and Fisheries reveals that China's 2,900 distant water vessels continue to plunder the ocean, often disguising their true location, employing destructive fishing techniques, and flouting territorial borders of sovereign nations.[6]

Vietnam's Fisheries Resources Surveillance allocates resources to address compliance with fishers and marine environmental protection. A governmental agency non-military task force established in 2013, it is responsible for patrolling, checking, controlling and detecting violators.[7]

Vietnam's activities in the protection of fisheries were highlighted in 2016 when they responded, along with marine scientists and government officials, to Ha Tin province where Taiwan's Formosa Plastics Corp. released chemicals, including cyanide and phenols, into the coastal waters, killing massive numbers of fish and ending many livelihoods in the fishing industry.

Fishermen are also cognizant that they are responsible in contributing to plastic debris accumulation in marine and coastal zones. Across the board, offshore fishing and aquaculture-related operations have also been identified as significant sources of plastic pollution into ocean basins and coastal ecosystems. Damaged fishing nets and abandoned, lost or discarded fishing gear (ALDFG) can be left offshore by fishermen.[8]

Vietnamese citizens love their East Sea and their fishermen. Their emotional and national connection goes well beyond any established regional bodies for managing fisheries in the sea. Oversight bodies like the Southeast Asian Fisheries Development Center and the Regional Plan of Action to Promote Responsible Fishing Practices (RPOA) only play an advisory role. Their non-binding language have failed to garner the people's hearts and minds about stewardship of marine resources. The locals only want their fishermen to follow the fish over international law.

10

America and the Constitution of the Oceans

Dark clouds hang over the horizon. The mariner weather lore, "red sky in the morning, sailors take warning," resonates sharply due to China's high-pressure drilling operations located in Vanguard Bank, inside Vietnam's exclusive economic zone (EEZ).

At the Center for Strategic and International Studies (CSIS) in Washington DC, academics, policy experts, State Department officials, and media convened inside the two-story glass façade headquarters to discuss this South China Sea issue at their annually held program.

In a past forum held in 2019, experts gathered to address why the international community, especially the United States, should speak out about China's seabed surveys in violation of the United Nations Conventions on the Law of the Sea (UNCLOS). It's the primary instrument governing the protection of seas. Initially adopted at the 1982 UN Conference on the Law of the Sea, it came into force after protracted negotiations in 1994. It has been widely designated as the "constitution for the seas" since it represents the most comprehensive international treaty ever concluded. The charter established rules for all types of use: navigation, fishing, oil and gas extraction, seabed mining, marine conservation, and marine scientific research.[1]

Bonnie Glaser, former senior adviser for Asia and director of the China Power Project at CSIS, claims, "If there's no response to these violations of the UNCLOS, it demonstrates that Beijing can violate international law with impunity."

China's violation of UNCLOS is evident in its indifference to the Arbitral Tribunal award at The Hague, its frenetic atoll building spree, not to mention the militarization of the Spratly Islands, fishing bans in disputed waters, and rampant ecological destruction of coral reefs. Based on its actions, Beijing's endgame amounts to a global security threat.

On the surface, US policymakers and Indo-Pacific experts see an approaching geopolitical storm, or rather more like a fast-moving typhoon, stretching across the South China Sea.

China's grab of atolls, rocks, and islands in the South China Sea remains highly contested among their neighbors. Of course, resource extractions, including that of oil and gas reserves, remain profitable enough to raise tensions between states.[2]

In recent years, there have been soft diplomacy measures undertaken by China and Vietnam to decrease tension, including joint coast guard patrols in the Gulf of Tonkin and the two Vietnam Navy warships attending a fleet review in China to mark the 70th anniversary of the People's Liberation Army Navy. But these gestures have also been overshadowed by Beijing's past actions of cutting underwater cables, live-fire exercises, and both nations' aggressive appetite for hydrocarbon and marine resources in the contested South China Sea.

China's incursion into Vietnamese sovereign zones, has spilled over into fishing. The collapse of fish stocks in the South China Sea continues to pose serious security and economic concerns.[3] The decrease in fisheries and changes in marine species migratory patterns force fishermen to go beyond their legal sovereign limits as they cast long and illegal lines into disputed zones.[4]

In the hot summer of 1974, the second session of the third UN Conference on the Law of the Sea was obscured amidst the fallout of the Vietnam War. Although the war had ended for America the year prior, the emotional and psychological toll at home had just begun. By 1982, the newly framed constitution of the oceans contained over 320 articles and at least nine annexes. Today, more than 157 nations have signed the treaty, but not the United States.

During the UNCLOS negotiations, most of the major distant water fishing countries have accepted the idea of an exclusive economic zone (EEZ), an area that stretches 200 nautical miles from the coast of a state in which fish stocks would be managed by that state. The new Biden administration may have its stars in alignment to encourage Congress to ratify the treaty, since the US does not want to undermine its value-based foreign policy nor global leadership.

According to James Kraska, international maritime law professor at the US Naval War College, "most ocean activities are located in EEZs, which encompass 36 percent of the total area of the sea. In fact, 90 percent of commercially exploitable fish stocks are located in the zones because the richest phytoplankton pastures lie within 200 miles of the continental masses."[5]

From many conversations with US policy experts, they all agree that unlawful and sweeping maritime claims in the South China Sea pose a serious threat to the freedom of the seas, including the freedoms of navigation and overflight, free trade and unimpeded commerce, and freedom of economic opportunity for the region's littoral nations.

During the Trump administration, the failure of the Senate Foreign Relations Committee to get the South China Sea and East China Sea Sanctions Bill, sponsored by Republican Senator Marco Rubio of Florida and Democrat Senator Ben Cardin of Maryland, out of committee and to the Senate floor was glaring. The bill's purpose was to punish Beijing for its "illegitimate" actions to claim territorial rights in the waters off of the country's coastline.

US forces operate in the South China Sea on a daily basis, as they have for more than a century. They continue to routinely operate in close coordination with like-minded allies and partners who share America's commitment to uphold a free and open international order that promotes security and prosperity.

China needs to become a responsible stakeholder in the contested region, but so does America. Currently, the US is bridging two vital areas—the Arctic and the South China Sea. In order to address and contest Russia's claims in the Arctic and those of China in the South China Sea, Washington should be a party to UNCLOS.

In 2021, we can expect to see more standoffs between Chinese and Vietnamese coast guard vessels over intractable territorial claims, with potential hydrocarbons and marine resources at stake.

The South China Sea should be considered central in the overall US-China relationship. The region is a hot zone for national security and heightened freedom of navigation exercises. Under the new Biden administration, it's unclear how Asia or the Indo-Pacific region will play in their foreign policy directives. Of course, the focus will still remain on the trade imbalance and intellectual property theft. But what about Washington's defense alliance with the Philippines and support for Vietnam's claims in the Paracel and Spratly Islands?

President-elect Joe Biden did reveal in a statement to the general press, "We're a Pacific power, and we'll stand with friends and allies to advance our shared prosperity, security, and values in the Asia-Pacific region." The soundest argument for ratification of the UNCLOS treaty is that the US does not cede any ground to national sovereignty but rather it expands American power by having a permanent seat on the International Seabed Authority (ISA), an autonomous organization established under UNCLOS in 1982 and the 1994 Agreement relating to Part XI of UNCLOS that encompasses what is referred to as the Area, and its resources, the "common heritage of man," or about 54 percent of the world's oceans.

Although short of declaration of alignment with any littoral states, the US State Department had gone on record to say in 2020 that "China's repeated provocative actions aimed at the offshore oil and gas development of other claimant states threaten regional energy security and undermines the free and open Indo-Pacific energy market."

Vietnam, a former enemy and now a US strategic partner, should remind Washington that Chinese pressure on Hanoi resulted in the suspension of an offshore natural gas project by Repsol, a Spanish firm, in Vietnam's own EEZ.

With the new administration, US policy makers may want to begin to organize an exploratory ocean policy team with the Association of Southeast Asian Nations (ASEAN) and China to establish a joint development area in the Spratly Islands, aimed at exploration of hydrocarbon reserves.

Of course, the US Navy does maintain its plans to exercise its Freedom of Navigation Operations by dispatching warships in the region. It's encouraging that Biden's newly appointed Secretary of State Anthony Blinken has said on record that "the US will show up and engage ASEAN on critical issues of common interest." It seems clear that America does intend to play a stabilizing role in the South China Sea as China seeks to impose its hegemonic control of the region and beyond.

According to the US, neither UNCLOS nor international custom negates the rights of states to conduct lawful maritime exercises, including military activities, in an EEZ of a coastal state without notice or its consent.[6] China, however, insists on its claim of sovereignty over the majority of the South China Sea, and argues that any activities undertaken in the SCS without prior notification violate its domestic and international law.[7]

Anders Corr, principal of Corr Analytics, believes that the current situation is a perfect opportunity for Washington to defend the integrity of the

EEZ principle, to draw Vietnam away from China and closer to the US, and to deny China access to hydrocarbons in the region.

If Vietnam stands down to China, it will signal a death knell for any other future major bidder for offshore drilling projects. ExxonMobil likely will also withdraw its Blue Whale, the multi-billion-dollar integrated gas-for-power development project.

US Senator Ben Cardin, a democrat from Maryland and ranking member of the Senate Committee on Foreign Relations and its Subcommittee on East Asia and the Pacific, continues to argue that America's failure to join the UNCLOS agreement enables China to continue to parade their disregard for international law in the disputed South China Sea. He has gone on record in stating, "[America's] failure to ratify the treaty also undermines our ability to fully work with our allies and partners in the South China Sea region. If we are not party to UNCLOS, it is difficult for the United States to rely on the treaty to determine the legal entitlements of mid-ocean features, which claims are lawful, and what exactly constitute the high seas."

After all, the US played an instrumental role in forming UNCLOS in the 1970s. In subsequent negotiations, it worked to modify the treaty language to assure that US national interests were safeguarded under UNCLOS. Unfortunately, when President Ronald Reagan came to office in 1981, he suspended America's participation in the tenth Session of the third Conference on UNCLOS. The suspension was largely because of US objections to the treaty's provisions on deep seabed mining.

A similar scenario may soon play out in the Arctic, thus there's pressure on the United States to ratify UNCLOS sooner rather than later. The freedom of the oceans of the world and coastal waters has surely proven to be a contentious issue in international law for the past four hundred years. The most influential and noted argument in favor of open navigation, trade, and fishing was that espoused by politician, shipwreck survivor, and Dutch jurist, Hugo Grotius in these words, "Let no man possess what belongs to every man."[8]

Grotius' views were largely shaped by Dutch ambitions in the East Indies trade in the seventeenth century. At the time, Spain claimed the Pacific and Gulf of Mexico, and Portugal the Atlantic south of Morocco and the Indian Ocean. The Dutch were an ascending international trading nation, who would soon have a worldwide empire enforced by their own naval military might.

Today, it's generally accepted that nations have a common interest to ensure maritime security in order to reach three complementary objectives: to facilitate the vibrant maritime commerce that underpins economic security, to protect the world's ocean resources, and to prevent maritime terrorism, hostilities, and crimes.[9]

In the Arctic, climate change is transforming the ocean in new and dramatic ways. The Arctic nations—Canada, Denmark, Finland, Iceland, Norway, Russia, Sweden, and the United States—have joined on the Arctic Council with other nations that want to have a piece of the natural resources and ship lane access that has now revealed itself in the unprecedented impact of global warming. This is also a global problem for all countries. The authors of new study in *Nature* estimate that sea level will rise 25–30 feet if the Greenland Ice Sheet disintegrates.[10]

Bold US leadership is urgently needed to rein in climate change and to protect valuable marine and coastal environments in the Arctic. "This has been an unprecedented year for people and planet. The Covid-19 pandemic has disrupted lives worldwide. At the same time, the heating of our planet and climate change disruption has continued apace. Record heat, ice loss, wildfires, floods and droughts continue to worsen, affecting communities, nations and economies around the world," claims Antonio Guterres, Secretary-General of the United Nations.[11]

With the increasing openness of the Arctic region for economic and military expansion, and according to Representative Joe Courtney, chairman of the House Armed Services Subcommittee on Seapower and Projection Forces, "there's little time to waste to ensure the US can approach any future discussions from a legitimate position based firmly in our ratification of the Law of the Sea." If the US does not ratify UNCLOS, it will be left behind in the race for the Arctic's vast natural resources. All arctic nations—Canada, Denmark, Norway, Russia—except for the United States, have ratified the Convention and have already submitted proposed limits for their extended continental shelves.[12]

The White House has plenty of domestic pandemic-related issues to deal with at the moment, a streamlined vaccine rollout, a stimulus boost for the economy, inequities in health-care, racism and criminal justice. However, climate change, the environment, and the protection of oceans remain pressing matters too.

11

Washington Slow to Connect the Blue Dot Network

Amidst Covid-19 and the escalating Cold War tensions between Beijing and Washington, it's no wonder that the US-initiated Blue Dot Network (BDN) remains stalled.[1]

BDN was conceived as an American counterpart to China's massive Belt and Road Initiative (BRI), when America, Australia and Japan reached an agreement to build infrastructure projects in the Indo-Pacific on the sidelines of the 35th ASEAN Summit that was held in November 2019. At this Indo-Pacific meeting, one of the most important and complex issues facing the China-ASEAN relationship, namely the South China Sea, was masked by the two sides putting on a game face and hailing together "the progress of the substantive negotiations" for the conclusion of the Code of Conduct (CoC) within the agreed framework of three years. The refrain remains the same: both ASEAN and the China need to cooperate for peace and stability in the region.[2]

This private-sector-focused and government-supported certification scheme is based entirely on quality infrastructure standards as set out in the G20 Principles for Quality and Infrastructure Investment. Kaush Arha, a senior fellow at the Atlantic Council has been vigilantly shepherding this proposed program. "BDN represents another chapter for US leadership to address global needs and safeguard the international system from predatory behavior."

Although President Trump was a no-show at the 2019 ASEAN summit, he sent National Security Advisor Robert O'Brien as his special envoy. In the eyes of the ASEAN leadership from Brunei, Cambodia, Indonesia, Laos, Malaysia, Myanmar, the Philippines, Singapore, Thailand, and Vietnam, the White House's decision earlier at the start of the Trump administration to withdraw from the Trans-Pacific Partnership only served to distance America from its Asian allies, just as China's influence rises in the region. Over the past four years, Washington's long-term allies have been deeply concerned about Trump's transactional approach to foreign policy.

First announced in 2013, the BRI was the cornerstone of Xi Jinping's signature foreign policy project. In 2017, it was written into the constitution of the Chinese Communist Party and 138 countries and 30 international organizations agreed to participate, with announced investments to link Asia, Africa, and Europe. The "One Belt One Road (OBOR)" is Beijing's umbrella term to describe a variety of initiatives, many of which appear designed to reshape international norms, standards, and networks to advance Chinese global interests and vision, while also serving its domestic economic objectives. As possible pushback to China's aspirational global recognition the Blue Dot Network emphasizes good governance at the epicenter of a Free and Open Indo-Pacific and that is the only true compass.

Many observers see fundamental problems associated with Beijing's BRI. There are suspicions about China's real intentions and criticisms against the poor quality of the infrastructure built by Chinese companies. According to an article in *World Politics Review*, there are tangible issues such as waste, political cronyism, and a slew of overambitious infrastructure projects that have resulted in unsustainable levels of debt for partner countries. In 2019, Malaysia, Myanmar, Nepal, and Pakistan have all cut back on their BRI projects out of fear of expensive contracts for potential albatrosses and mounting debt.[3]

The irony of the BDN is that America's own aging infrastructure is crumbling. Far too many bridges are unsafe. According to the American Road & Transportation Builders Association (ARTBA), the number of structurally deficient bridges in the US in 2018, if placed end-to-end, would stretch 1,216 miles—the distance between Miami and New York City. The costs are estimated to be around $3.8 trillion, states the World Bank.[4]

No one denies that the world has plenty of underdeveloped countries that are still in dire need of basic infrastructure investments. During this pandemic, there's even a greater downturn on essential transportation needs and connectivity with worldwide trade routes. "The Covid-19 pandemic has gravely wounded the world economy with serious consequences impacting all communities and individuals. Moving rapidly across borders, along the principal arteries of the global economy, the spread of the virus has exploited the underlying interconnectedness—and vulnerabilities—of globalization, catapulting a global health crisis into a global economic shock that has hit the most vulnerable the hardest," says Mukhisa Kituvi, Secretary General of the United Nations Conference on Trade and Development (UNCTAD).

Bridging the poor countries' connectivity requirements will require a resolve to fund the Blue Dot Network. Since most regard this as mere rhetorical to thwart China's juggernaut Belt and Road Initiative. It's no wonder that even the choice of "blue" sharply contrasts China's "red." The certification aims to encourage private sector-led projects to adopt stronger labor protection and be sustainable, both financially and environmentally. At a Center for Strategic & International Studies (CSIS) program held in 2020, experts acknowledged that the certification process will not come cheap. It must be rigorous enough to persuade private sector investors to put their money into riskier places. Creating a multidimensional quality standard will require negotiation among engineers, builders, architects, procurement officials, financiers, rating agencies, and perhaps other stakeholders.[5]

Unfortunately, many of China's BRI projects fail to live up to the standard of fair and good governance practices. Chinese construction efforts are marked by poor quality, corruption, environmental degradation, and a lack of transparency and community involvement. It's also difficult to reconcile Beijing's claims to promote "green development" when it is the world's largest greenhouse gas emitter, plastics debris polluter, and illegal, unreported and unregulated fishing nation over the past decade.

This lack of stewardship is seen in the dams constructed along the Mekong River that are cutting off some fish species from their upriver breeding grounds. Furthermore, the dams already built upstream in China's section of the Mekong River have seriously reduced water flow in downstream Southeast Asian countries.

At the 2019 ASEAN Summit, National Security Advisor O'Brien was quick to criticize China. "The region has no interest in a new imperial era where a big country can rule others on the theory that might makes right. America is helping our ASEAN friends uphold their sovereignty," he said.

As a result, only three ASEAN heads of state from Thailand, Vietnam, and Laos participated in the meeting. Trump's lack of interest in ASEAN was evident throughout his presidency. Nevertheless, the chair's statement reflected the emphasis on reinforcing ASEAN-US strategic cooperation.[6]

The BDN is projected as a multi-stakeholder project that brings together governments, the private sector, and civil society to insure high-quality, trusted standards for global infrastructure in an open and inclusive framework. Fortunately, there's increasing support among Southeast Asian leaders to support the 1982 UNCLOS, that it should be the basis of sovereign rights and entitlements in the disputed South China Sea. Also, US leaders fully endorse quality infrastructure projects in Southeast Asia.

The Blue Dot Network positions itself as an entity that desires to normalize quality standards applicable to infrastructure investments. It remains unknown in what substantive way that this network will boost a globalized quality standard.

"Through the Blue Dot Network, the United States is proud to join key partners to fully unlock the power of quality infrastructure to foster unprecedented opportunity, progress and stability," claims David Bohigian, former acting president and CEO of the United States Overseas Private Investment Corporation (OPIC), an agency which recently morphed into the US International Development Finance Corporation (USDFC) and doubled the size of investment capabilities from $30 billion to $60 billion.[7]

Analysts have suggested that the BDN is about controlling regional power since it reflects a shared Indo-Pacific strategy among nations that want to offset China's rising power. Beijing's trillion-dollar BRI has been characterized as "debt-trap diplomacy" and the US has repeatedly warned that Chinese telecoms giant Huawei's 5G expansion posed a national security threat to Washington.

While the trade battle lines drawn between Washington and Beijing, Mike Pompeo, former Secretary of State, made it clear that the US "wants a transparent, competitive, market-driven system that is mutually beneficial for all involved." His comment offers a non-diplomatic rebuke to China's state-led economic model, rife with corruption and without transparency.

Although Beijing's signature multi-trillion-dollar global economic outreach grows, some developing countries are saying no to BRI financing. This is increasingly so in ASEAN countries like Malaysia, Myanmar, and Vietnam, who are the most concerned about the high costs of Sino-led projects, increased Chinese influence, and undue leverage over their sovereign interests.

Since the BDN encourages the adoption of trusted standards for "quality global infrastructure development in an open and inclusive framework," it offers an excellent opportunity for Vietnam, a country unwilling to look to China for all of its infrastructure needs—roads, rail, airports, ports, telecoms, and water. According to the World Bank, Vietnam projects at least $25 billion needed for infrastructure investment per year to maintain its economic growth, which has fallen to 2.9 percent since the pandemic.[8]

With a relatively high public debt burden and nascent capital markets, Vietnam has been searching for alternative ways to meet its infrastructure needs and the BDN could be poised to fill that need.

While Hanoi has endorsed the BRI and the Asian Infrastructure Investment Bank, it remains very skeptical of the strategic implications of Chinese investments and that includes what other ASEAN neighbors have experienced in falling into a "debt diplomacy trap." Many policy experts portray the China's BRI as a calculated geopolitical strategy that entangles countries in unsustainable debt and gives China undue influence.[9]

Yet, the global economic recovery from Covid-19 increases the need for lending for many countries.

Many organizations, politicians, and researchers in the US and Europe have been claiming that countries participating in the BRI could become economically and politically dependent on China since they would become Beijing's debtors, even lose natural resources and be subject to Chinese espionage.[10] For some in Washington, the Blue Dot Network's sole purpose is part of a US broader messaging plan to persuade developing countries in Asia to not rely on Chinese infrastructure funds.[11]

Chinese contractors have a poor track record in Vietnam and there's increasing protests since many Chinese projects have resulted in costly delays, cost overruns, and shoddy construction quality. The Cat Linh-Ha Dong metro line in Hanoi, funded by Chinese loans and built by a Chinese contractor, was originally scheduled for completion in 2013 but remains unfinished. The project cost has also doubled, from $377 million to $771 million.

"As a developing market, [Vietnam's] growth cannot be maintained without sustainable infrastructure. Overall, it is projected that Vietnam will require $605 billion in infrastructure spending by 2040," says Dr. Oliver Massmann, an international attorney in Vietnam.

Vietnam hopes to attract large amounts of infrastructure development funds through foreign direct investment for seaport upgrades and expansion.

"Since the US-Japan-Australia Blue Dot Network aims to encourage private sector investment in approved projects, it follows that Vietnam would encourage investment from the United States," claims Carl Thayer, emeritus professor at the University of New South Wales Canberra at the Australian Defense Force Academy.

Vietnam's priorities would be to overcome the shortage of infrastructure funding for its priority projects, and to ensure that Vietnam does not become too dependent on one source. The Vietnamese government would welcome founding under the BDN, not so much as a counter to China's BRI, but as an addition to it.

"In the final analysis, BDN infrastructure funding will depend on resuming economic growth and being able to pay off Vietnam's growing debt," adds Thayer.

The BDN was designed to offer alternatives to China's BRI funding. It should be viewed as a major geo-diplomatic initiative to counter China's influence. But the BDN cannot compete with the amount of funds China is willing to throw at infrastructure development.

David Dodwell writes that China has grand ambitions for the BRI.[12] To date, more than 60 countries—accounting for two-thirds of the world's population—have signed on to projects or indicated an interest in doing so. Also, Morgan Stanley has predicted China's overall expenses over the life of the BRI could reach over $1.2 trillion by 2027.

The BDN is best seen as part of the broader Washington-led efforts to persuade developing countries in Asia to not rely entirely on Chinese funds for infrastructure. Furthermore, with the Biden administration's Build Back Better for the World initiative announced at the 2021 G7 Summit, the timing could not be better to push forward in conjuction with the BDN.

This global pandemic has caused downward economic spirals and increased debt burdens for developing countries. It may prove useful for Southeast Asia to take a page from America's Benjamin Franklin's fabled *Poor Richard's Almanack*, "rather go to bed without dinner than to rise in debt," especially if China is holding the note.

12

Red Flags in the Pacific Islands[1]

Tony Yao used to fish in his outrigger canoe the coastal waters of Tahiti. But the decline in fish populations has forced his family to move, in order to find less exploited fishing grounds.

His story is just one of many from the South Pacific that hint at a paradise being lost. The coral and volcanic archipelagos scattered across this massive expanse of water face a number of threats, but climate change and overfishing are arguably the most serious.

Figures from the World Bank show that nearly one-third of global fish stocks are overexploited.[2] Commercial fishing is taking place at biologically unsustainable levels, driven by the rising global demand for seafood. Poor catches in China's coastal waters have driven Chinese fishermen to the farthest reaches of the Pacific to meet the country's growing demand for seafood, especially tuna. Because foreign fleets dominate offshore fishing with catches transshipped for processing outside the region, Pacific islanders derive little benefit.

The Pacific tuna fishing grounds are the largest in the world, accounting for more than 60 percent of the global catch. Tuna, a $3 billion a year industry, remains a vital source of food and employment for islanders, and fish has historically been their only major renewable resource. The fisheries access fees paid by foreign fishing vessels have become a significant source of revenue for local governments in many islands in the Pacific. China, by far, dominates tuna fishing in the open ocean outside the Pacific nations' exclusive economic zones (EEZs).

Oceanic fisheries target four species of tuna—skipjack, yellowfin, bigeye, and South Pacific albacore. Supplies, however, are overexploited and conservationists say urgent action is needed to ensure that the tuna fishery remains viable.

"Under a high emissions scenario, ocean-warming and acidification are expected to reduce live coral reef cover by 50 to 75 per cent. Models also indicate that a significant shift in the distribution of all tuna stocks in the Pacific will occur by 2050," according to Valerie Allain, a fisheries research scientist in Noumea, New Caledonia.

The Western & Central Pacific Fisheries Commission (WCPFC), often called the "Tuna Commission," oversees an international convention among Pacific nations and their fishing grounds.[3] The commission aims to ensure rules are fair for all foreign nations operating in EEZs—up to 200 nautical miles off shore—of the Pacific nations as well as in the high seas, or international waters, between latitudes 20N and 20S.

The latest WCPFC figures show that around 3.24 million metric tons of tuna were caught across the Pacific in 2017. Of that, 78 percent (nearly 2.54 million metric tons) came from the area managed by the commission. Global tuna catches in 2017 totaled almost 4.72 million tons. However, these catch rates threaten to drive some of the tuna species to extinction. For instance, in Fiji, land degradation, reef destruction, increased pollution, and the unchecked exploitation of marine resources are responsible for the decline in coastal fishing.

The presence of extensive areas of international waters among the EEZs complicates the region's fishery protection efforts. It is here that China dominates with more than 669 vessels out of a total of 1,300 foreign-operated fishing vessels casting their nets into Pacific waters. Their fleet is enabled by government subsidies on fuel and shipbuilding, which assist new enterprises and allow less efficient ones to keep operating.

Mounting evidence suggests that the WCPFC needs to do much more in addressing fishing sustainability. Pew Charitable Trusts' assessment concluded that with respect to individual species management, the WCPFC record is mixed.[4] To be fair, the Commission has made efforts to combat IUU fishing by adopting a Compliance Monitoring Mechanism on port state measures and requiring the use of unique vessel identification numbers. The latter includes a requirement that smaller vessels fishing outside their state's EEZs display International Maritime Organization numbers.

James Movick, director general of the Pacific Islands Forum Fishery Agency, claims that, "The biggest single threat to the tuna fishery is the lack of control of mainly foreign vessels operating on the high seas."[5] "When fishing in our 200-mile EEZs, they are subject to regulation and a robust fishery management regime; outside, it is pretty much a free for all—and tuna do not recognize the boundaries of our EEZs," he adds.

Also, the Pacific Islands Tuna Industry Association in Fiji maintains a registry of all fishing vessels licensed to fish in the region.[6] John Maefiti, an executive officer of the association, indicates that there are 627 Chinese fishing vessels registered, and the majority of them are long liners. He adds that "there are some Chinese-owned vessels that fly Pacific Islands flags, and the second biggest player is the Republic of China Taipei fleet."

More than 90 percent of China's distant water fishing (DWF) vessels fly the Chinese flag. This is because Chinese regulations of its DWF actions are notoriously less vigorous than the regulation of its own domestic fisheries. In a 2020 Overseas Development Institute report on Chinese DWF fleets, the research effectively stated that "as a flag state, China does not have a particularly strong record of engaging with the international community and complying with regional fisheries management organizations (RFMOs)." They are international bodies made up of countries that share a practical and or financial interest in managing and conserving fish stocks in a particular region.

Cecile Matai, responsible for offshore, coastal, and lagoon fishing license applications in Papeete, the Tahitian capital of French Polynesia, remarked: "No Chinese ship has a fishing license, but there are Chinese ships in our ports for refueling and obtaining supplies, or with mechanical problems which are being repaired."

Matai suggests that these undocumented vessels contribute to IUU fishing. Because of the dominant role of state-owned enterprises in the Chinese economy, China continues to offer government subsidies for fuel and shipbuilding to assist new and existing enterprises to dominate tuna fishing. Throughout Pacific island countries, there's a lack of data transparency and public reporting on IUU fishing.

Professor Daniel Pauly, a noted marine biologist and global fisheries expert at the University of British Columbia Institute for the Oceans and Fisheries, and other researchers claim in their research paper, "China's distant water fisheries in the 21st century," that "while Chinese DWF in a given EEZ may be perfectly legal (i.e. with complete access agreement negotiated between China and the host country), they may remain 'unreported.'"[7]

Climate Change Impact on Fisheries

According to the International Union for Conservation of Nature (IUCN), climate change is compounding food security issues in the Pacific islands, with harvests from fisheries expected to fall between 10 to 30 percent by 2050 as water temperatures rise.[8] Rising water temperatures, lower levels of oxygen, and shifting ocean currents are already affecting the four main tuna species, as well as fish habitats, food webs, fish populations, and fishery productivity.

Climate change is also leading to more extreme weather in the South Pacific. The 2015–16 tropical cyclone season was one of the most disastrous on record. Cyclone Winston, which smashed into Fiji in February 2016, was the strongest ever to make landfall in the southern hemisphere. The National Oceanic and Atmospheric Administration (NOAA) confirms that as the sea surface temperatures become warmer, hurricanes, typhoons, and tropical cyclones become more powerful.[9]

The UN Intergovernmental Panel on Climate Change expects sea levels to climb by 83 centimeters by the end of this century if greenhouse gas emissions remain high. The levels are creeping up even faster in the Pacific, where at least eight low-lying islands have been submerged in recent years.

Marine scientists are now finding links between ocean acidification and a decline in tuna populations. A new study in *Science* outlines the impacts warming waters have on commercially important fish species. Changes in the ocean temperature are affecting fragile ecosystem food webs. "While tuna spend their time in open water, tuna species as well as tuna fisheries depend on healthy coastal ecosystems such as coral reefs," states UN coral expert, Jerker Tamelander.

As climate change effects strengthen, small island states are seeking other economic lifelines aside from foreign aid. However, development prospects remain poor. In the past five years, China has exercised greater influence in the region through political and diplomatic engagements, significant aid provisions, and strategic trade and economic ties.

China's parallel actions reflect, on the one hand, that they are providing grants and loans to many hardscrabble island communities, and on the other, they are scooping up as much fish as possible in the Pacific. This action has been precipitated by the Parties to the Nauru Agreement (PNA) that comprises the Federated States of Micronesia, Kiribati, the Marshall Islands, Nauru, Palau, Papua New Guinea, the Solomon Islands, and Tuvalu. The eight countries, plus Tokelau, represent the largest sustainable tuna purse seine fishery in the world, because they control most of the waters in which tuna are found. For skipjack, most commonly used for canned tuna, around 50 percent of its global supply is harvested in the waters of the PNA members.[10]

The economic stakes associated with tuna and the PNA are high since revenue has risen from $60 million in 2010 to $500 million in 2016.

However, fishing nations are not pleased with the sharp increase in fishing license fees. This year in French Polynesia, there's been a wave of online protests and petitions to ban Chinese tuna fishing vessels.

China's highly subsidized fishing fleets have been buying up the biggest share of licenses. As a result, competing fishers have seen declining catches, and depressed tuna prices are pushing local tuna fleets to leave family-run fishing enterprises. Now, there are Chinese fleets expanding into offshore fisheries, including in the Western Central Pacific Ocean.

Geopolitical Shifts in Pacific

In terms of diplomatic and security engagement, China has actively ramped up its presence in the Pacific through participation in regional organizations, and highly publicized visits. An examination of trade, investment, aid, and tourism data shows that China is rapidly becoming a dominant economic force in the region, well ahead of the United States. This "debt book diplomacy" effort often translates into island-wide political influence and even strategic equities.[11]

Chinese aid to the region now exceeds that from New Zealand, Japan, and the EU, making Beijing the third largest donor in the Pacific Islands. In Micronesia, the Compact of Free Association with the US comes to an end in 2023. "There's a sense of desperation if US funds dry up," claims Mae Bruton-Adams, an environmental and sustainability consultant on Pohnpei.

Chinese investment has brought some new opportunities for regional economic development. With demand for seafood around the world now outstripping supply from wild fisheries, sustainable aquaculture is one industry that may develop in the Pacific. One planned Chinese project of $320 million is Tahiti Nui Ocean Foods fish farm on the Hao atoll; approximately 920 kilometers east of Tahiti.

China has during the last decade moved quickly to emerge as the third largest donor to the Pacific Islands. Its estimated annual aid is somewhere between $100 million and $150 million, claims Jian Zhang in *China's Role in the Pacific Islands Region*.[12]

A recent US National Defense Strategy paper warns that China is effectively employing "predatory economics" to achieve a re-ordering of the Indo-Pacific region to China's advantage and to tamp down any support for Taiwan.[13]

"Pacific Ocean-based fishing and tourism provide almost $3.3 billion to the national economics of Pacific countries and territories," according to the Secretariat of the Pacific Regional Environment (SPREP). However, Pacific islands are vulnerable to rising seas and the impact of climate change on coral reefs, fisheries, and resources.

In the South Pacific, Tony Yao and his ancestors have maintained a deep connection to the marine environment. After all, the ocean has been the provider of life itself. Though having faced extraordinary challenges before, island cultures are now experiencing ones they could not have imagined, including unprecedented sea levels, storm surges from tropical cyclones, ocean warming, acidification, disappearing coral reefs, and competition from registered and unregistered fishing vessels.

13

Post Hague Decision Offers Ecological Security

Few policymakers or writers could have predicted that the Philippines would win in the landmark ruling from the Permanent Court of Arbitration (PCA) in The Hague, against China's claim and seizure of the Scarborough Shoal in the South China Sea. It constituted a huge win for Manila over the reef lying about 140 miles from the Philippines coast, which China occupied in 2013. Nearly five years after the arbitration award, the situation in the disputed region remains tense and maritime incidents persist.

The five-judge tribunal's unanimous decision packed in the 501-page document hit China broadside like a typhoon and was nothing short of an unequivocal rejection of Chinese claims in the contested sea. Of course, the hardest blow for Beijing was the Tribunal's conclusion that "there was no legal basis for China to claim historical rights to resources within the sea areas falling within the 'nine-dash line.'"

It's true that the award did not make pronouncements on the numerous territorial disputes that exist with respect to islands, rocks, reefs, and shoals as they were questions were beyond the jurisdictional reach of the Tribunal.[1]

The timeliness of the court's decision created a policy buzz around a myriad of issues, from militarization, capacity building, legal challenges, enforcement, and ecological security at a discussion held at the Center for Strategic & International Studies (CSIS) during the annual South China Sea conference.

I listened to keynote speaker Daniel J. Kritenbrink, a senior director for Asian affairs at the National Security Council, who served in the Obama administration, as he reiterated Washington's recognition of how China has been challenging the international order established by the United States more than 70 years ago. He was making a compelling case for America's continued commitment to freedom of navigation in and overflight above the South China Sea. Kritenbrink stated that "Washington had no need or interest in stirring tension in the South China Sea" as any pretext for involvement in the region. However, he did add, that the "US will not turn a blind eye to this important waterway in return for cooperation elsewhere in the world."

Of course, observers and panelists knew in advance that China would summarily reject the PCA ruling, but there were many questions about what's next, including an ASEAN response, a response that would certainly be tested at future ASEAN foreign ministers' meetings.

The Hague's watershed ruling provided much more than background noise for the gathering of ASEAN ministers. It was forcing members to choose between China and the US. The CSIS program offered both a prologue and potential recommendations during the scheduled ministerial-level meetings. What's clear is that both ASEAN and the US remain at best skeptical of China's propaganda and public promises of non-hegemonic ambitions. They are not blind to China's security concerns, including the need to feed its people.

It's important to understand how the Tribunal's past ruling impacts today's geopolitical maneuvers in the South China Sea. There's little doubt that at US Indo-Pacific headquarters, strategic, operational and tactical teams are putting together new evolving paradigms to respond to China's aggressive actions in the disputed waters.

Here are the highlights of the Tribunal's rulings:

Historic Rights and "Nine-Dash Line"

The Tribunal concluded that, to the extent China has historic rights to resources in the waters of the South China Sea, such rights were extinguished to the extent they were incompatible with the exclusive economic zones provided for in the Convention. The Tribunal also noted that, although Chinese navigator and fishermen, as well as those of other States, had historically made use of the island in the South China Sea, there was no evidence that China had historically exercised exclusive control over the waters or their resources.

The Tribunal concluded that there was no legal basis for China to claim historic rights to resources within the sea areas falling within the "nine-dash line."

Status of Features

The Tribunal ruled that none of the Spratly Islands are capable of generating extended maritime zones. The decision also held that the Spratly Islands cannot generate maritime zones collectively as a unit. The court found that none of the features claimed by China was capable of generating an exclusive economic zone. Also, they ruled that it could—without delimiting a boundary—declare that certain sea areas are within the EEZ of the Philippines, because those areas are not overlapped by any possible entitlement of China.

Lawfulness of Chinese Actions

The Tribunal considered next the lawfulness of Chinese actions in the South China Sea and ruled that certain areas are within the EEZ of the Philippines. As a result, the court ruled that China had violated the Philippines' sovereign rights in its EEZ by (a) interfering with Filipino fishing and petroleum exploration, (b) constructing artificial islands, and (c) failing to prevent Chinese fishermen from fishing in the zone.

Harm to Marine Environment

The Tribunal considered the effects on the marine environment of China's large-scale land reclamation and construction of artificial islands at seven features in the Spratly Islands and found that China has caused severe harm to the coral reef environment and violated its obligation to preserve and protect fragile ecosystems and the habitat of depleted, threatened, or endangered species.

Despite Chinese assertiveness in the South China Sea, this is the first time Beijing had to face the international justice system over its claims in the disputed sea. For more than a decade, Chinese scholars have published articles on state-owned media outlets such as *China Daily*, *People's Daily*, and *Xinhua* and science journals to promote Beijing's "nine-dash line." Their efforts have only served to generate pushbacks from other claimant nations, including Brunei, Malaysia, the Philippines, Taiwan and Vietnam.

It's noteworthy that the *People's Daily* focuses on themes such as China being attacked in its traditional fishing area, and China disagreeing with American and Japanese accusations of self-isolation and use of force.[2]

"The Chinese government has repeatedly said that it will not recognize any ruling. But it has also worked hard since Manila brought the case in early 2013 to get the Filipino government to drop it. That is because being branded an international outlaw will involve significant reputational costs for Beijing. It will undermine China's narrative that it is a responsible rising power that deserves a greater hand in global governance," stated, Greg Poling, director of the Asia Maritime Transparency Initiative at CSIS.

Although the environmental panel was the last presentation of the CSIS program, there's increasing attention on the region's environmental insecurity, including degradation in coastal breeding habitats, coral reef destruction, climate change, biodiversity loss, and fisheries depletion.

Professor John McManus from the Department of Marine Biology and Ecology at the University of Miami, an internationally recognized coral reef specialist, presented a case for an international peace park in the contested Spratlys. His passionate presentation outlined how the region is headed towards a major fishery collapse if there is no consensus reached over a freeze on claims and no joint resource management.

He shared with me and others over lunch how in February 2016, he and a film crew arrived in the Spratly Islands to document changes in the marine life. His dive revealed dead coral and few fish. As a former Peace Corp volunteer assigned to the Philippines, he conducted coral reef surveys and along the way, met his future wife, Liana Talaue-McManus, a marine scientist, who also shares his passion for the tiny brilliant cathedrals beneath the sea.

"The ability of coral reefs to exist in balanced harmony with other competing and limiting biological agents has been severely challenged in the last several decades," claims McManus. Southeast Asia is the center of biodiversity for the reefs of the world and contains about one-third of the world's coral reefs.[3]

After completing his PhD in biological oceanography at the University of Rhode Island in 1985, McManus relocated to Bolinao on Luzon, the largest island in the Philippines, to commence his focused studies on coral reefs and fisheries.

He and Liana have witnessed firsthand the incredible diversity of organisms associated with coral reefs since they are the repositories of great promise for improving the quality of life for the world.

The ebullient broad shouldered senior marine scientist knows that excessive sedimentation from deforestation, construction, dredging, mining, and other sources cuts back on coral growth and settlement and often results in the burial of coral communities, particularly small non-reef communities.[4]

McManus understands that marine ecosystems, particularly coral reefs, are affected by climate change. For example, the Great Barrier Reef in Australia has lost over 30 percent of its coral cover due to mass coral bleaching event attributed to abnormally high sea surface temperatures.[5]

Food security and renewable fish resource challenges are quickly becoming a hardscrabble reality. With a dwindling fish catch in the region's coastal areas, fishing state subsidies, overlapping claims of EEZs, and mega-commercial fishing trawlers competing in a multi-billion-dollar industry, the decline of fish is one of the issues at the heart of this sea of troubles.

The South China Sea is truly the ecological heart of Southeast Asia. In fact, a study as early as 2004 by the UN Environment Program cited the high concentration of coral reefs in the region, representing nearly 34 percent in the world, despite occupying only 2.5 percent of the total ocean surface.

The UN Environmental Program confirms that the South China Sea accounts for as much as one tenth of global fish catch, and that China will represent 40 percent of global fish consumption by 2030. Overfishing and widespread damage to coral reefs necessitate a science-driven policy intervention to safeguard this vital sea.[6]

"We are headed toward a major fishery collapse and this environmental catastrophe will impact hundreds of millions of lives; it's time to act now," exclaimed McManus.

The Spratly Islands' immense biodiversity cannot be overlooked. The impact of continuous coastal development, escalating reclamation, and increased maritime traffic in the sea calls for marine scientists and policy strategists to study the sustainability of the ecological seascape and to navigate the development of science diplomacy.

During the environmental panel session, McManus's powerpoint slides, along with that of fellow marine scientists, Dr. Edgardo Gomez from the Philippines and Dr. Kwang-Tsao Shao from Taiwan, bolstered the urgency to establish a marine peace park, or at the very least, to foster a sweeping conservation movement among all the region's fishery associations. "The term peace park does not necessarily imply that it is sited within an area in conflict, although the term does indicate a propensity for this kind of protected area to reduce violent conflict and bring more harmony to a region."[7]

The Convention on Biological Diversity targets the establishment of 10 percent of marine protected area coverage. With regard to the Spratlys, transboundary protected area arrangements have often been proposed. There is a well-established precedent for these, although they are primarily in the form of parks on land. In 1988, the Commission on National Parks and Protected Areas of the International Union for the Conservation of Nature (IUCN) listed 70 protected areas in 65 countries that straddle borders.[8]

Among policy experts whom I spoke to during this program, they seem to agree that subsequent ASEAN meetings may offer a clear opportunity and diplomatic roadmap for a unified response to The Hague ruling that includes a commitment to the Code of Conduct, a respect for international law, and a peaceful resolution of disputes.

Increasing numbers of marine scientists from claimant nations and ASEAN agree that there's a fishery collapse unfolding, which creates a more than sufficient common ground to address food security in the region. Looking at the next several years, rising demand from growing populations and economies is in a direct collision course with over exploitation, pollution, habitat destruction, and climate change.

Furthermore, the number of coral reef and fish species in these contested waters has declined precipitously from 460 to around 261 species, and the list of critically endangered species now includes Green turtles, giant clams, and Hawksbill turtles.

Drs. McManus, Gomez, and Shao all point out that many of the coral reef fisheries along the coasts of the South China Sea have been heavily overfished, especially along the coasts of southern China, Vietnam, Malaysia, and the Philippines.[9]

The evidence shows that harvests of adult fish have been in a steady and steep decline. Marine science studies suggest that offshore reefs may be critical to preventing local extinctions of targeted fish species.

Dr. Gomez, an eminent marine scientist from the University of the Philippines-Marine Sciences Institute (UP-MSI), presented abundant evidence that China's construction over the rocks and shallow reefs should be an alarm for the world. Chinese reclamation of these atolls and islands involves dredging sand from the ocean floor and dumping them over entire coral reef systems.

During the course of writing this book, Dr. Gomez died on December 1, 2019. He was a recognized voice in the protection of the Philippine archipelago's vast marine resources and led the world's first national-scale assessment of damage to coral reefs leading to international conservation initiatives such as the replanting of corals.

The future of Philippine policy towards the maritime row and attitudes towards the arbitration award remain unpredictable.

Meanwhile, ASEAN knows that the South China Sea links their global economies, since it remains an energy shipping route and provides the essential sea lanes between Southeast Asian islands. With increasing clashes among fishing vessels in the contested region, the regional leadership may be taking notes on how to reduce fishing incidents rather than to resorting to sending more fleets into the commons.

Washington policy observers recognize ASEAN leadership's lack of a united position on the international tribunal's ruling. What's certain is that the ongoing dispute continues to damage the unity of the trading bloc and remains a test for the ASEAN Political Security Community (APSC), and the China factor is the most critical concern.[10]

Although the Philippines versus China award has had little to none enforcement outcomes, it does offer a klieg light on China's environmental degradation in the South China Sea.

Perhaps a few of the suggested confidence building measures below might be taken up by ASEAN leaders at their future scheduled meetings. After all, the ASEAN region is a major contributor to global biodiversity, containing four of the world's 34 biodiversity hotspots and three mega-diverse nations. The region's biodiversity and ecosystems are essential to the economic, social, and environmental well-being of the millions of people, to the growth of agricultural export economies, food security, and livelihoods.

Recommendations:

- Expand science cooperation among all ASEAN marine scientists.
- Establish complete freedom of scientific investigation in the contested atolls and reclaimed islands.
- Place aside all territorial claims.
- Draw upon the shared expertise of the Asian Fisheries Society.
- Renew the operations of the ASEAN Institute for Peace and Reconciliation (APIR).
- Create a regional Marine Science Council to address environmental degradation issues.

- Foster dialogue for a proposed marine peace park.
- Propose a science-led committee to examine the Antarctica Treaty as a possible model for the South China Sea.

ASEAN recognizes the importance of fisheries to food security and to the economy. Because China has become the world's top producer and exporter of fishery products. It should be held responsible to operate in alignment with shared sustainable practices among its neighbors.

If there are to be any fish left in the contested sea, an ASEAN ecological agreement—led by Brunei, Malaysia, the Philippines, Indonesia, and Vietnam—can steer others to unite around a proposed international peace park or a cooperative marine protected area situated prominently in the Spratlys. It's the first step in fostering trust and confidence among neighbors and in implementing a common conservation policy.

14

Social Media Invites Waves of Environmental Nationalism

Stroll around any of Vietnam's major universities from Hanoi to the Mekong Delta and ask students about what concerns them, and they are almost uniform in their responses—the country's environmental emergencies, from degradation, loss of habitats and biodiversity, air and water pollution.

Vietnam's fast-track economic growth over the past several decades arrived at the expense of the environment, leading to polluted waterways, extensive loss of wildlife, marine biodiversity, and a near fisheries collapse. A global environmental performance ranking places Vietnam in 141st place out of 180 economies. The index measures what countries are doing to protect the environment and human health from environmental damage. Vietnam was given a rating of 170 for air quality.

Although its air pollution is not as severe as that in many Chinese cities, the increased number of motorbikes in both Saigon (Ho Chi Minh City) and Hanoi is all too evident to locals and visitors. Stepping off the curb on Dong Khoi street in Saigon is downright dangerous with more than 7.8 million motorbikes emitting harmful carbon monoxide.

Even prior to Covid-19, Hanoians wore masks to curb respiratory ailments since there are nearly 6 million motorbikes in the city of 8 million—compared with around 700,000 cars. Tourists and locals are quick to reach for their smartphones for ride-hailing apps such as Grab for on-demand transport service. But while these motorbikes remain the capitol's main source of transportation, the government plans to ban motorbikes in downtown areas by 2030 in an effort to ease air pollution and congestion in the city.

But what struck me the most over the past decade is the rising tide of environmentalism. Since 1986, the country has embraced an ambitious program of economic reform known as *doi moi* or renovation. From the early 1980s, Vietnam was a one of the world's poorest country following decades of war against the French and America. The country's remarkable journey from low to middle-income status lifted more than 40 million people out of poverty. As a reporter in the country from 1998 until now, I have witnessed how the country leaped into the international system by joining the World Trade Organization in 2006, and celebrated in 2020, the 25th anniversary of normalization of relations with the US, a former enemy and now friend and trading partner.

Vietnam's flood of foreign investments includes US multinationals like Nike, which employs over 500,000 young women in its factories. However, the country's "jump into the ocean," a metaphor my Vietnamese friends use to refer to their joining the international system, has come with the required responsibilities of compliance, efficiency and transparency but the challenges remain.

The Vietnamese state or Communist Party has consistently pledged to balance ecological improvements as part of their march towards continued economic growth, but there remain dramatic failures wrecking the pact between the economy and environment.

The international community is cognizant that Vietnam has taken aim at some of the environmental challenges reflected in policy initiatives: the Vietnam Climate Change Strategy (CCS 2011), Vietnam Green Growth Strategy (GGS 2012–2020), National Strategy for Environmental Protection (NSEP 2012–2020), Vietnam REDD + program (2012–), and a range of legal instruments on environmental, forest and water protection.

Nevertheless, these bold environmental decrees are always hampered by unsatisfactory policy implementation, and reflects a top-down approach which generally fails to connect with the communities at risk and thus obstructs the development of a functional environmental state.[1]

In countless conversations with local academics, and reporters, I have learned how Vietnam's national identity is closely interwoven with their relationship to the East Sea and especially towards their fishermen. It holds them in a net of community, culture, and heritage. Thus, it's natural to understand that the Vietnamese want to protect what they love. In many conversations with fishermen, they seem to echo what the poet Pablo Neruda wrote, "I need the sea because it teaches me."

Somehow, it's as if America's great national treasure, the scientist-poet Rachel Carson is reaching out to Vietnam's new transnational generation and declaring that the future of their nation is at stake. Her inspiring 1962 book, *Silent Spring*, revealed a culture-crashing exposure of chemical pollutants and the impact on Earth's ecosystems and it created a tidal wave of support and understanding about the ecological connections between nature and society.[2]

While most Vietnamese students are not familiar with her work, they inform and educate the public effectively through millions of Instagram messages about environmental disaster associated with the poisoning of their environment.

The blog posts and images of dead fish washed up along Vietnam's Central coastline caught the attention of millions in Vietnam in April 2016. This massive fish kill was caused by water pollution from the toxic industrial waste discharged by a steel plant of Formosa Plastics.

Vietnam has a population of 98 million, in which increasing numbers are internet users. Social media has proven to be the first choice of activism tools among the educated Vietnamese youth. Their online activism was highlighted in Ha Tinh province, the environmental crime scene of the Formosa Plastics Corporation's steel plant, where citizens took to the streets, even rallying near Hoan Kiem Lake in Hanoi, to protest the Taiwanese corporation's improper discharge of untreated toxic wastewater into the East Sea.

It was a turning point for Vietnam's environmental and social activists, many who became first time citizen reporters and posted their photos of tons of dead fish swept along their 125-mile coastline, devastating sea life and local economies dependent on fishing and tourism. With more than 45 million users, Facebook rapidly became a lightning rod for environmental activism and sparking demands for improved environmental safeguards.

For a while, internet censors blocked Facebook over the environmental hashtag, #IChooseFish protests or the Vietnamese version, *#toichonca*. By owning and operating all media agencies and outlets, the Party ensures that it has an unimpeded ability to censor all forms of social and political expression.

The Formosa environmental disaster succeeded in creating a nation-wide opposition to unchecked industrialization in unprecedented ways. Citizens, especially the youth are demanding a clean environment.

Environmental activist Nguyen Huynh Thuat believes that the Formosa steel plant ecological disaster "generated a social protest against an unchecked industrial development and signaled a new wave of environmentalism through Facebook postings."

"It is largely through social media that young people like Hoang Thi Minh, the leader of CHANGE (Center of Hands-on Actions and Networking for Growth and Environment), and 360 organizations, based in Ho Chi Minh City, that are bringing people together to protest ecological damage to Vietnam's landscape and seacoast," claims Le Thu Mach, a Lecturer at Ho Chi Minh National Academy of Politics.[3]

Dr. Mach's insightful and timely PhD dissertation, "Emerging Social Media and the Green Public Sphere in Vietnam," completed at Monash University in Australia, encouraged me to reach out to her, who helped frame part of this chapter. She successfully investigated how journalists and others have been using social media to translate public opinions into environmental policy changes.[4]

While she quickly acknowledges that the Formosa waste water discharge was serious and resulted in 100 tons of destroyed fish and marine species, the overall fishing capacity of Vietnam is between 3.1 to 3.2 million tons a year, according to the Ministry of Agriculture and Rural Development (MARD). Thus, although the loss was devastating to the livelihoods of citizens in Ha Tinh province, represents just 0.003 percent of Vietnam's total annual fishing capacity. In Mach's analysis of the social media coverage, she believes that some activists on social media posted fake photos of the dead fish taken outside of Vietnam.

"Since journalists were prevented from gaining access to the shore site, they lacked the photographic evidence and, as a result, too many newspapers used illustrated photos from the internet, which showed massive death of fish but not from Vietnam," claims the academic.

For example, the photo published in *Nha bao va Cong luan* (Journalists and Public Opinion), a newspaper by the Vietnam Journalists Association, on April 28, 2016, was actually a photo of fish death in Tianjin, China. In fact, if one conducts a search today on the Formosa sea disaster, one would see that most photos have already been removed.

Mach says, "Journalism is excluded from any truth-seeking practice." Her argument rings true with my own experiences in reporting in Vietnam over the past several post-*doi moi* decades.

I agree with her that Hanoi's media gatekeepers prohibit the flow of independent news coverage. In my reporting especially between 1997 and 2000 in Hanoi, I was closely monitored by Public Security, but eventually became friends with my minders and the two-person detail security team following me around the city as I interviewed locals about the impact of the internet on their lives.

In the fall of 1999, I entered the country as an invited research fellow at the Hanoi School of Business through a cooperative program with Dartmouth's Tuck School of Business. My research focus was on the internet and its impact on media development. Dr. Truong Gia Binh, the founder and executive chairman of FPT Corporation, one of the largest internet service providers in Vietnam, but at the time we met, he was the dean of the business school, and also happened to be the son-in-law of General Vo Nguyen Giap, the most decorated military leader, who defeated both France and the US during the Vietnam wars. Now divorced from the late General's daughter, he has remarried and is regarded as one of the nation's richest businessmen.

Upon entering the Vietnam National University campus on a crisp and clear November day, I was immediately informed that Dean Binh demanded to see me right away. Apparently, before my arrival on campus, Vietnam's public security had paid a courtesy visit to the Dean and was more than inquisitive about how an American journalist from *The Washington Times* had received this open-ended academic research invitation.

"Mr. Borton, I am not really sure how you arranged this research appointment but let me be clear, you have a visa that enables you to conduct interviews and research in the country but please do not come back to this campus," exclaimed Binh. It's evident there was a showdown between Vietnam's Public Security and the politically connected dean, and I was directly caught in the cross-hairs of public security and nascent steps of integration with the West. I left his office and never returned during my extended six-month guest stay in Vietnam.

As a result, and into early 2000, I conducted numerous interviews with young Vietnamese seated in the newly opened internet cafes and also met with Mr. Nguyen Anh Tuan, the founder of *VietnamNet*, the most popular online news publication in the country. In fact, he even offered me an opportunity to advise his company and staffers just when the internet was rapidly being adopted by users. There were nearly 200 young staffers and many of them were bilingual recent college graduates. His website was attracting nearly 4 million users a week and is owned by the state. Despite this, I was surprised that this youthful Communist Party member, did not hesitate to push for transparency and competition in the selection of political leadership.

Tuan was also recognized as the founder of VietNet, the first internet service provider in Vietnam, an achievement for which we was selected as one of the 10 outstanding young Vietnamese by the prime minister in 1996. In March 2011, Tuan informed me that he was resigning as editor of the popular online newspaper. A few years earlier the management of the paper was transferred to the Ministry of Information and Communication. His decision was based on his new appointment as a distinguished fellow at Harvard's Shorenstein Center on Media, Politics and Public Policy at the John F. Kennedy School of Government. He subsequently co-founded the Michael Dukakis Institute for Leadership and Innovation and the Boston Global Forum. It's clear that much credit must be paid to Tuan for paving the way for the country's current social media developments.

Social media first arrived in Vietnam with the *Yahoo!360* blog service in June 2005. *Yahoo!360* was nominated by *VietnamNet* as one of the top 10 trends in ICT in Vietnam in 2006.

According to my source Professor Mach, *Yahoo!360* blog service was closed down in December 2009 due largely to techno-social issues. After 2009, social media users migrated to other platforms such as Yahoo Plus, Opera, Multiply, Wordpress or Blogspot. It was in 2013, that Facebook became a prominent social media platform in Vietnam. However, the service was periodically blocked by Hanoi's internet gatekeepers. By 2018, Facebook had garnered almost 61 percent of users. But with the passage of the Cyber Security Law in June 2018, Hanoi purposefully attempted to divert traffic away from the site to local or 'Made in Vietnam social media platforms such as Zalo (60 million users), Mocha (8.7 million), Gapo (2.6 million) Lotus and even the Communist Party of Vietnam Propaganda Committee site, VCNet. Most of the local portals have failed to find traction among Vietnam's youthful user base.[5]

| JAN 2021 | **VIETNAM** THE ESSENTIAL HEADLINE DATA YOU NEED TO UNDERSTAND THE STATE OF MOBILE, INTERNET AND SOCIAL MEDIA USE | ★ VIETNAM |

TOTAL POPULATION	MOBILE PHONE	INTERNET USERS	ACTIVE SOCIAL MEDIA USERS
97.75 MILLION	**154.4** MILLION	**68.72** MILLION	**72.00** MILLION
URBANIZATION: 37.7%	VS. POPULATION: 157.9%	PENETRATION: 70.3%	PENETRATION: 73.7%

Source: *Datareportal and VNetwork.*[6]

Vietnam has a dismal record for free media, ranking 175 out of 180 countries, where the ranking of 1 is regarded as the freest, based on an annual index compiled by media watchdog Reporters Without Borders. Observers and rights activists believe that the state's stranglehold on the news media and social media appears to be tightening rather than relaxing.

In the fall 2020, Vietnam's journalists and social media users faced a new obstacle to independent reporting through another government media censorship decree issued by Prime Minister Nguyen Xuan Phuc, in which he ordered a penalty for "posting or disseminating information unsuitable for the interests of the nation and people" that carries an administrative fine of up to 200 million Vietnamese dong ($8,600). Additionally, the offense of "providing untrue information to the public that distorts and defames individuals and organizations" carries fines up to 40 million Vietnamese dong ($1,723).

At this time, police in Vietnam arrested a well-known Facebook user, Truong Chau Danh, 38, a former journalist, over allegations of abusing democratic freedom and publishing posts against the state according to a recent Ho Chi Minh City police newspaper. Danh owned a Facebook page with nearly 168,000 followers.

In the case of environmental reporting, journalism is the de facto voice of the authority and is the recognized instrument to promote its political agenda and to reinforce social and political ideologies. The Vietnamese Communist Party operates a complex mechanism of censorship that streamlines all media content produced in the country. It's becoming clear that increasing ranks of news consumers find the news coverage within state newspapers to be inadequate, and they are turning to alternative sources of news to learn more about politically sensitive issues that have been censored.

In my communication with Professor Mach, she said that when journalism is excluded from the truth-seeking practice, the citizen media or social media emerges, sometimes as the single source for the event.

In her comprehensive dissertation research and interviews, she included a quote from Mr. Truong Minh Tuan, former minister of information and communication from a press conference held on June 30, 2016. "The Party and State do not have any policy to conceal the truth. It's not only the people, but also the Party and State need to know the truth. It requires all ministries and local government to involve seeking the truth."

The then minister added that "not long after the [Formosa] incident, to make more favorable conditions for the investigation process, we requested Vietnam[ese] journalists to adhere to the Press Law, reduce the volume, temporarily cease the journalistic investigation, stop any inference, and wait for the conclusion from the authorities." The former minister is now in jail on corruption charges.

In two decades, Vietnam's rapid industrialization has reduced poverty and helped raise the country's per capita income to over $3,600, eclipsing that of its poorer neighbors Laos and Cambodia. However, this rapid build-at-all-costs industrialization occurs generally at the expense of the environment, generating pressures on natural resources.

Despite an early government policy discussion in 1985 that called for a national conservation strategy, it was never officially adopted. In 1991 Hanoi announced a national plan for the sustainable management of the country's natural resources. However, it was neither institutionalized nor implemented until the Ministry of Science, Technology and Environment (MOSTE) and the Ministry of Natural Resources and Environment (MONRE) were created, along with the Vietnamese Environment Agency in 2008.

Pressure from the international donor community and local shareholders resulted in some successful policy changes, but the new environmental laws failed to provide any legal enforcement of compliance. In fact, five years ago the Environmental Performance Index listed Vietnam in the top ten worst countries for air pollution.

Because of Vietnam's environmental enforcement weakness, a litany of damages to the land and water continues to mount: two thirds of Vietnam's forests are in decline mainly due to massive illegal logging; air pollution increases daily due to the rising number of motorbikes; wastewater is released untreated; and industrialization pollutes rivers and streams.

It's no wonder that Vietnam has witnessed the emergence of prominent non-governmental organizations like People and Nature Reconciliation (Pan Nature), Centre for Water Resources Conservation and Development (WARECOD), ECO Vietnam Group, Green Innovation and Development Centre, Save Vietnam's Wildlife (SVW), and the Mekong Environment Forum (MEF). In a disclaimer, I am a co-founder of MEF along with Nguyen Minh Quang, a Geography Lecturer at Can Tho University, where we produced numerous citizen science workshops to help local volunteers address environmental issues.

There are increasing numbers of young Vietnamese networking for environmental activism, many of them stridently opposed to marine plastics pollution and coal plants. I have seen firsthand scores of Vietnamese volunteers participate in CHANGE. It featured popular Vietnamese actors, musicians, and artists wearing gas masks, performing before a devastating backdrop of smog and climate change devastation.

Fortunately, the space for civil society protests against environmental threats continues to open up. I was in Hanoi where I saw the Green Trees Movement unfold. This broad-based citizen-led movement protested against Hanoi government's arbitrary decision to cut down thousands of large old trees lining the city's streets.

In 2015 on a March spring day, hundreds of Vietnamese gathered on the streets of Hanoi and displayed red banners written with yellow-colored slogan, "Greet the Party, Greet Spring." The Vietnamese simply could not understand why the authorities intended to chop down 6,708 ancient trees firmly planted on perhaps as many as 190 streets in historic Hanoi. The Green Trees movement mobilized a public outcry through a Facebook page, especially among young citizens. However, the protests failed to save nearly 2,000 trees from being chopped down.

According to Professor Ngoc Anh Vu from the Social & Policy Sciences at the University of Bath, "These trees, making up more than a quarter of the total city greenery, would be cut down or replaced for reasons such as the trees were dying, decayed, or bent, or posing risks to road users during rainy seasons, or different kinds of trees were planted on the same street creating a poor aesthetic outlook (implying that the government wanted uniform trees), or that trees were detrimental to planned infrastructure projects."[7]

On the Facebook page, organizers claimed that they had sent representatives to attend a meeting with the Hanoi People's Committee to discuss this destruction of Hanoi's trees on May 6, 2015, and another meeting on May 8 with Hanoi's members of Parliament.

The Tree Movement has to be considered a success since most of the trees were protected. The end result was increased confidence in meetings held with government officials and it gave citizens a stronger voice through crowdsourcing. Finally, it proved to the authorities that protests can be done in an orderly and peaceful manner and reflected a bottom-up victory for people-led democracy and for upholding the role of social networks.[8]

After the Formosa issue, Green Trees published a book, *An Overview of the Vietnam Sea Disaster*. Until March 2019, Green Trees consistently targeted the political goals through an environment lens. They also successfully produced a documentary about sustainability and green issues in Vietnam. More young Vietnamese are using their social media sites to broadcast to the public alternative views on the protection of the environment. These tree activists directly challenged the government's authority and can claim success for their green campaign in saving some of the trees.

Green, the designated color or symbol for environmental issues, now conveys a new meaning and contrasts with Red, the color for Communism in Vietnam.

Grassroots environmental movement participation and exposure to scientific research in general continue to increase among the 40 million plus net-savvy users. It's part of what I have witnessed and also promoted as the co-founder of the MEF in Can Tho, Vietnam. Our workshops are laying the roots for what may be called citizen science or civic science.

Citizen science is the collaboration between scientists and interested local citizens to broaden the scope of research and help compile scientific data through community-based monitoring and internet-driven crowdsourcing strategies. The increasing prevalence of internet and smartphone usage throughout Vietnam offers the necessary digital infrastructure for doing so. MEF's co-founder Quang has been engaged in reaching out to more student volunteers to gain their interest in helping local farmers address the dramatic climate change impacts on their livelihoods. Our NGO continues to produce environmental workshops at Can Tho University and in the field.

A central feature of citizen science has been a shift in the way scientific concepts and information are communicated to non-experts. It's a shift not just in how the media is being used but also in media content, with hard data being supplemented by anecdote and narrative—for example in the form of blog posts.

As Richard Louv explains in his 2011 book, *The Nature Principle: Human Restoration and the End of Nature*, "Citizen scientists collect more than data. They gather meaning."[9]

At the Wilson Center in Washington, Anne Bowser has championed a Facebook Citizen Science initiative that has been working with the United Nations on the Sustainable Development Goals.[10] Along with SciStarter.org and their efforts to offer collaborating data exchanges for discovery of global projects, there are increasing opportunities for young Vietnamese and others in the region to become champions for environmentalism.

Some recent examples include a Facebook page that successfully connected the medical community and health experts with the public about news on the transmission of Covid-19.[11] It is an excellent illustration of how Facebook was effective in dispelling fake news since it drew together the authoritative health experts with the general public about the news about Covid-19 pandemic. Another example is a Facebook page of the fact-checkers community "Chung tay chống tin giaả," meaning countering fake news.[12]

As part of the attention to the increasing ranks of online environmentalists, Mekong Matters Journalism has assisted with the professional training of citizen reporters in the Mekong with their well-placed network of eyes on the myriad of environmental challenges in the Mekong Delta. This network was established in September 2014 and continues to link environmental bloggers, citizen reporters, and journalists from Cambodia, China, Laos, Myanmar, Thailand, and Vietnam.

According to Adam Hunt, a former InterNews media trainer, who has led environmental media training programs in Southeast Asia, including several in Can Tho, Vietnam, believes that "the rapid pace of development in the Mekong is creating more opportunities for reporters to find innovative and important stories." Hunt's Mekong Eye website regularly showcases news from these reporters and the fluid site has become an effective information resource for villages and government officials.

Over the course of the past five years, the Vietnamese have continued to protest numerous other environmental issues, from textile companies, sand dredging, the hydropower project in Cat Tien National Park, to pig farms, coal power plants, and cement factories.

"Social media is proving its important role in promoting environmental awareness to the young generation in Vietnam and throughout Southeast Asia", says Khiem Nguyen, founder of Mekong Delta Youth, who's now pursuing a Master of Resource Management in New Zealand.

This ecological awareness and the use of data are essential in shaping stories that the Hanoi censors now welcome since they go against the grain of "fake news" and enable a public sphere that opens up something new and unique in Vietnam—a participatory culture. During the present health storm, when reliable news is more important than ever, citizen science is helping fight misinformation and biased news coverage. This so-called rise of "infodemic," that refers to the rapid spread of misinformation about vaccines and their efficacy, becomes as dangerous as the virus itself.[13]

It seems more than possible that Vietnam's policy officials will recognize that through citizen science workshops, young trained volunteers will prove instrumental in facilitating the much required collaboration in communicating about the current pandemic and related environmental security issues.

It's simple. A citizen-driven culture means that large numbers of Vietnamese farmers, fishers, and educated youth are realizing their capacity to produce and share media with each other, often responding critically to the products of the state-controlled media. Their adoption of social media facilitates a free and fluid circulation of their observations across a range of social media platforms.

This era of social media contributes to impactful environmental protests that have proven effective in bringing together local citizens to protest against environmental injustice. But the jury is still out if it will lead to a structural change in Hanoi's environmental policy making and implementation or in press freedom.

15

Fishing Frontlines: A Sea Change for Southeast Asia Fish Exports

The tides wait for no one. Southeast Asia fishers live this truth daily and know that their catches are in decline, particularly for those hardscrabble lives casting their nets close to shore.

The perils for these fishermen are well-documented and include clashes with other commercial trawlers, resource depletion, water pollution, and limitations in catch traceability, along with dangerous labor conditions.

The trade war between the US and China is also harpooning the Asia-Pacific region and fueling sea changes in fish wars. By sheer volume, more than 50 percent of the world's catch of marine and river fish, and almost 90 percent of global aquaculture come from this area.

To be clear, Asia's fisheries and shrimp farms are essential for the food security and livelihoods of rural and coastal populations. However, the distribution imbalance of global fish production, coupled with fish trade tariffs, challenges socio-economic growth in poorer countries. According to the Asia Foundation, nearly 64 percent of the fisheries' resource base is at medium to high risk from overfishing, with Cambodia and the Philippines among the most heavily affected.[1]

The destructive fishing practices include poison fishing, the indiscriminate use of sodium cyanide that stuns fish, and blast fishing with dynamite or grenades. The nearshore stocks are increasingly overexploited by artisanal fishers.

The Global International Waters Assessment (GIWA) report on the South China Sea found blast and poison fishing to be major problems throughout Southeast Asia.

This fishing picture is exacerbated in the South China Sea where coastal fishing grounds have been depleted of nearly 30 percent of their unexploited stocks. Consistent with this trend, unsustainable fishing practices have been confirmed in local areas within the Vietnamese EEZ, where overexploitation can be largely attributed to the incursion of foreign fishing vessels.

While there are some past attempts to establish cooperative mechanisms to frame fishing issues in the South China Sea, most of them are designed to protect marine biodiversity and not the fishermen. These include the Sino-Vietnamese agreements on fishing in the Tonkin Gulf, the UNEP/GEF South China Sea project, the Marine Stewardship Council, the Sulu-Sulawesi marine ecosystem, and the State-market-NGO programs.[2]

For emerging economies like Vietnam, the export of frozen catfish, or *tra*, has been a long-standing contentious trade issue in the US market. Despite the US having imposed anti-dumping duties now calculated at 25.39 percent, *tra* fillets accounted for 90 percent of the total imports to the US market in the first half of the year.

The growing popularity of fish in global markets like the EU, the US, China, and Japan frays more than the nets of fishers—it also disrupts and denigrates the marine commons. This "tragedy of the commons," first coined by scientist Garret Hardin in 1968, describes what happens when groups or nations, pursue their self-interests over what's best for all. Overfishing in the disputed South China Sea is an example where weak fishing regulations contribute to rampant illegal practices that often go unreported.

In the South China Sea, or the East Sea as the Vietnamese prefer to call this body of contested water, it's clear that the ocean's resources need more marine stewardship. Total fish stocks in the South China Sea has been depleted by 70 to 95 percent since the 1950s, according to experts at the Asia Maritime Transparency Initiative at the Center for Strategic and International Studies in Washington. The region is facing the rising risk of catastrophic biodiversity losses. The wider environmental impact often gets intertwined in tricky sovereignty issues, as China's long-distance water vessels continue to ensure their fish catches are boundless. Chinese steel trawlers are pushing their boundaries daily and prowling the deep seas as far away as the coast of West Africa.

Unfortunately, Vietnam's climate change is having a deleterious impact on their fish stocks. The Notre Dame Global Adaptation index compares a country's level of vulnerability to climate change to its readiness to deal with these impacts. The science is clear: Vietnam is ranked as the 77th most vulnerable country and the 63rd least ready of 177 countries in the index to adapt to climate change. It's hard to overstate Vietnam's reliance on fisheries. A comprehensive study ranks Vietnam as the most sensitive and vulnerable country in the world.[3]

The good news is that Vietnam exhibits some moderate capacity to adapt to climate change given its status of economic development, and relatively good governance.[4]

The climate change situation in the US offers no relief model for Hanoi. The summer of 2021 has brought more wildfires, freak monsoons, heat-buckled roads, collapsed buildings, and algal booms or increased red tides in Florida destroying hundreds of tons of marine life.

In a review of the status of the world's fisheries and aquaculture, the Pew Charitable Trust has recommended to members of the Committee on Fisheries of the UN Food and Agriculture Organization (FAO) that they take measurable steps to improve fisheries governance.

It's no wonder that at this rate the Organization for Economic Cooperation and Development and the FAO forecast a fishing growth rate of a meager 1.2 percent over the next decade.[5]

"IUU fishing has negatively affected Southeast Asia since it costs the region billions of dollars annually, accounting for more than 2.5 million tons of fish a year, or as much as a third of the regional catch," writes Peter Chalk, an expert on maritime security at the US Naval Postgraduate School.[6]

Despite these troubling forecasts, the fisheries sector remains a cornerstone of the Vietnamese economy and has contributed to an average growth rate of 7.9 percent. However, Vietnam, like other ASEAN countries and China, also contributes to illegal and unreported catches, disrupts ocean beds and destroys coral.

According to the FAO, Thailand's fishing industry is also on a downward trend. The failure in stewardship of marine life falls heavily on an increasing number of marine biologists, who recognize that baselines have shifted and that governments always act in their own self-interest with little regard for future conservation and sustainability.

"Amidst overexploited fisheries and further climate related declines projected in tropical fisheries, marine-dependent small-scale fishers in Southeast Asia face an uncertain future," acknowledge Dr. Daniel Pauly and Lydia C.L. Ten, both of whom are scientists associated with the Institute for the Oceans and Fisheries at the University of British Columbia.[7]

European Union sources reveal that the Commission completed a formal visit to Vietnam from May 16 to 24, 2018 to evaluate the progress of Vietnamese authorities in addressing the deficiencies that led to the yellow card adopted last October. The European Commission has said that it will continue to cooperate with Vietnam to redress the situation that led to the issuance of the yellow card. Credit must be given to Vietnam's authorities since they amended their Fisheries Law to require all fishing vessels over 15 meters to be equipped with satellite positioning equipment, which came into effect in 2019. Additionally, Vietnam issued a white paper entitled "Combating Illegal, Unreported and Unregulated Fishing" in May 2019.

Furthermore, Vietnam's Prime Minister approved the establishment of the National Steering Committee on the Prevention and Control of Illegal Fishing. However, Vietnamese fishermen still clash with Indonesian and Malaysian law enforcement forces at sea.

According to Enrico Brivio, a spokesman for the EU's Environment, Maritime Affairs and Fisheries, Vietnam's current yellow card status remains under review. Thus, it joins the ranks of other violators including Thailand, The Philippines, and Papua New Guinea.

In 2020, Tran Dinh Luan, director ceneral of the General Department of Fisheries, reaffirmed that the "key is solving the problem of Vietnamese illegal fishing in foreign waters." Luan know that even if there is one violation the ability to withdraw the yellow card status remains very difficult.

Meanwhile, Vietnam's neighbor Cambodia remains on the export sidelines, having been issued a red card restricting all of its fish products into the EU. The country's Tonle Sap Lake, once rich in biodiversity and a source for fish protein for over three million people, is also under threat of collapse from climate change and upstream dam construction.

Southeast Asia fisheries are at a turning point and the declining fish catches and collapsing fisheries need to be recognized as a food security risk for the region's path to development. The Asia-Pacific Fishery Commission, a regional consultative forum, reaffirms that with over three million fishing vessels in the region and among that, two million fish in the South China Sea and Bay of Bengal, the future of the fisheries is in peril.

Over the past several years, there are more Vietnamese delegations led by Vice Minister Vu Van Tam and joined by senior officials from the Ministry of Agriculture and Rural Development (MARD) who have scheduled administrative reviews with the US Department of Commerce and International Trade

Administration officials, to respond to the imposed penalties and tariffs on their fish exports.

In 2019, the Vietnamese government agreed to join the Port State Measure Agreement (PSMA) established in 2016, to prevent and fight IUU fishing. Many observers believe that this adhesion to PSMA is an important first step in bringing Vietnam's fisheries legal framework in line with international standards. It's clear that IUU fishing is one of the greatest threats to the sustainability of global fisheries.

Article 119 in the UNCLOS identifies the sharing of data among signatory states as a priority. Available scientific information, catch and fishing effort statistics, and other data should be shared on a relevant basis through competent international organizations at sub-regional, regional or global levels, and with participation by all states concerned.[8]

It's also equally important to have more countries become signatories to the PSMA. This compulsory requirement to collect and share data on fishing vessels entering ports and to deny entry to vessels that have been fishing illegally needs ratification by important port and flag states around the world. The World Economic Forum reveals that 86 states have signed the agreement. While it's difficult to place a specific amount on the economic losses associated with IUU fishing, it may be somewhere between $10 billion and $23.5 billion.[9]

"Enforcing effective port controls as a region is equally critical: illegal operators always look for the path of least resistance, and will therefore make their way to any ports with lax controls," states PSMA expert Dawn Borg Costanzi.[10]

It's encouraging news that Vietnamese Prime Minister Nguyen Xuan Phuc has called for a national action plan to crack down on IUU issues, which would bring together 62 seafood companies to sign off on sustainable fishing practices.

With so much at stake, there seems to be increasing agreement among fishery experts and policymakers that fish stocks can be replenished through good management: enforcing quotas, cooperating on marine scientific research, penalizing transshipments, and developing marine protected areas to safeguard against further exploitation.

16

A Farewell to Arms in Vietnam[1]

The shower of bombs dropped by the US B-52s along the Ho Chi Minh Trail during the Vietnam war struck both fear and resolve into the heart of 18-year-old North Vietnamese Army veteran Khuat Quan Thuy.

The elaborate highway system, surfaced in places with crushed rocks and logs, ran through the jungles of Cambodia and Laos and was the primary artery for moving troops, supplies and vehicles through some of the world's harshest terrain.

It was under the jungle canopy and within hidden bunkers and caves that the Communists received instruction in Ernest Hemingway's *A Farewell in Arms* and *The Old Man and the Sea*.

"Before the war, we knew little about foreigners and had little access to books. Yet, when I was in school, I read *The Old Man and the Sea*," says Thuy, now a 70-year-old editor of *Van Nghe*, Vietnam's literary weekly newspaper. Indeed, it was not unusual for soldiers to gain an understanding about Americans through a reconnaissance of literature. Vietnam appears to have taken a page straight out of Washington's practice of cultural diplomacy and the Cold War cultural export of American Corners, once part of the United States Information Agency (USIA).[2] Over the decades, more Vietnamese friends have shared stories on how and why they studied Hemingway during and after their war experience.

There is a clear convergence of cultural diplomacy and the effectiveness of literature in establishing bridges between nations. Over 20 years ago, at a joint Vietnam-US program called "The Future of Relations Between Vietnam and the US," Vietnamese delegates expressed an eagerness to learn more about the United States and welcomed the opportunity to read more works by Hemingway and other great American authors.

At this Johns Hopkins School of Advanced International Studies (SAIS) event, General Le Van Cuong from the Vietnamese Ministry of Public Security, asserted that "Americans have not really understood Vietnamese people and we should put our past behind us." With the war over and history marching forward, the United States' relations with Vietnam have become deeper and more diverse since former US president Bill Clinton announced the formal normalization of relations in 1995.

As the only journalist present at the US-Vietnam program, I witnessed and participated in some surprising exchanges, including one with Dr. Dao Duy Quat of the Commission for Ideological and Cultural Affairs in Hanoi. During a discussion on educational exchanges, he exclaimed, "Really, our country is eager to know more about America and we welcome the opportunity to read more great American authors like Ernest Hemingway."[3]

I was intrigued by Quat's views on Hemingway and approached him and other Vietnamese delegates over the course of the program. As a student of American Studies, I thought his views, and perhaps those of other Vietnamese, might be attributed to the writer's simple style. With a literacy rate of 98 percent, it's no wonder that the promotion of learning and respect for teachers are part of the core values of the Vietnamese.

Ironically, the Vietnamese appreciation of Hemingway originated because the USIA forbade the distribution of his works through its centers during the Cold War. Officials deemed his work, especially *The Sun Also Rises*, *The Old Man and the Sea*, and *A Farewell to Arms*, as not being hostile enough to Communism. This irony would not have been lost on the writer who possessed a rich sense of humor and remains the most widely translated author in the world, awarded the Nobel Prize for Literature in 1954, and the subject of global scholarly meetings, workshops and review essays.

There are many reasons why the Vietnamese are reading Hemingway and it may have more to do with his craft than the author's politics. Carl Eby, a Hemingway scholar, suggests that "Hemingway's stylistic preference for simple, direct sentences, and his disdain for inflated diction make him ideal for second language learners."

Hemingway's lean, disciplined prose made writing and living seem simple. His spare sentences—with few adverbs or adjectives—make it easy for foreign readers to access his stories. Catherine Cole, a creative writing professor at the

University of Wollongong in New Zealand, suggests that Vietnamese audiences favor such word usage and the enigma and reflection in Hemingway's work.

Few readers dispute that *A Farewell to Arms* captures Hemingway's criticism of war, as the story of Frederick Henry and Catherine Barkley reveals lives caught up in the chaos and confusion of war. His novel illustrates the complexities of patriotism and unreliable international alliances.

Others believe there were many details of Hemingway's life, which attracted Vietnamese readers, including his participation in Spain's civil war, his close friendship with Cuba (an ally of Vietnam), and his suicide, which could have been used by North Vietnam as evidence that he was tired of US politics and society.

It was largely America's hubris and naiveté that caused it to get stuck in the tragic Vietnam quagmire. While it may not be feasible to rebrand or recreate the American Corners program, there is a need to revisit parts of it. It only takes a moment to examine how the former Trump administration's misguided foreign policy showed disdain for allies and embraced dictators like Kim Jong-un and Vladimir Putin. The America First campaign slogan set the United States on an isolationist path. It's therefore no wonder that some foreign leaders are reshaping alliances.

Ten years ago, the late US Senator Richard Lugar, a former chair of the Senate Foreign Relations Committee, stressed the need for the United States to interact with the world if it wished to change how other countries perceive it. Hemingway's work demonstrated his understanding of this. He was involved in three major conflicts, the two world wars and the Spanish Civil War, and wrote about each of them.

Loss was inevitable, and he knew firsthand that war claims lives, innocence and truth. These universal truths were also examined by Vietnam's Bao Ninh in his *The Sorrow of War*, in which he, like Hemingway, connects the tragedy of war to the loss of youth. Hemingway's *Soldier's Home* deals with the alienated soldier and his displacement from society.

For Vietnamese intellectuals and writers, South Vietnam's surrender on April 30, 1975 marks not just the fall of a country and the exodus of citizens but also the fall of literature. Saigon would be renamed Ho Chi Minh City, its boulevards, avenues, and streets also renamed to commemorate revolutionary figures, events, and slogans. So too, "Vietnamese literary history would eventually be dismantled, systematically re-written, or outright erased; books would be banned, confiscated, and burned; writers silenced, censored, and imprisoned," writes Hai-Dang Phan, author of *Reenactments*, who was born in Vietnam and now lives in Iowa City.[4]

It's noteworthy that the late Senator John McCain, himself a prisoner of war in Hanoi, recalled Hemingway's character Robert Jordan from *For Whom the Bell Tolls*. Like McCain, Jordan "knows that life is good, and it will be a very bad thing to lose his life. But he's very stoic about it."

At the end of the novel, Jordan faces his demise with a powerful reflection that McCain quoted before his own death. "The world is a fine place and worth the fighting for and I hate very much to leave it." Many scholars contend that war marked the loss of American innocence and the bell continues to toll for all who are engaged in combat.

Frank Stewart, the founding editor of the University of Hawaii journal *Manoa*, tells a story that after the war, a US veteran and a Vietnamese veteran meet and discuss their experiences. The Vietnamese veteran reveals that he likes American literature and he names a number of his favorite authors, such as Steinbeck and Hemingway. He tells the American veteran that his government sent troops into battle with copies of many American novels so that they could get to know the Americans better.

US Naval Academy historian Brian Van De Mark, author of *Road to Disaster: A New History of America's Descent into Vietnam*, identifies in his book a litany of American failures to know the enemy in the atmosphere of the Cold War and at the build-up toward US troop commitments in Vietnam.[5]

At the SAIS program in Washington, the Vietnamese delegation showcased its young and educated elites to the American audience. These included Nguyen Hai Yen and Le Linh Lan, two women from the Institute for International Relations who articulated, in excellent English, Vietnam's views on democracy, human rights and freedom of religion—and even of the press.

In 1997, I reported on Vietnam's renovation, referred to as *doi moi*, which began in 1986. Over the past several decades, the nation's literature has reflected changes in political liberalization, economic transformation, and globalization. Vietnam's writers previously adhered to the uniform Communist Party revolutionary culture of socialist realism with its collective ethos. It has now evolved into a pluralized culture that validates individual experiences.[6]

Many Vietnamese writers, such as Phan Hon Nhien and Vo Thi Hao, have been fellows at the International Writing Program at the University of Iowa. The Bureau of Educational and Cultural Affairs at the US State Department supports this cultural initiative.

This cultural diplomacy had an impact on several Vietnamese writers. Da Ngan is a 70-year-old author and former resistance fighter who wrote her career-defining novel *An Insignificant Life* about the travails of war, hardship, poverty and perseverance. She married fellow writer, Nguyen Quang Than.

Both were members of the Vietnam Writers' Association in Hanoi before relocating to Ho Chi Minh City in 2008.

She left her family to join the War at the age of 14. Her father was a member of the national resistance organization called the Viet Minh and was imprisoned in Con Son—notorious for its penal facilities during the French colonial era—where he died. In 1968 she went to Cau Mau province to write for newspapers where there was a library for "war warriors." "I was dazzled by the novels of Hemingway, John Steinbeck and Jack London," she says.

It was a cool autumn afternoon in Hanoi when I was introduced to Le Minh Khue, who fought against American troops from 1965 to 1969 and was also a war correspondent from 1969 to 1975. Her story, *The Distant Stars*, chronicles her military experience and is widely celebrated in Vietnam.[7]

"The day I went to the front, I left my family, my parents, brother, and sisters, this sweet home of mine shaken by the turmoil of war. My comrades and I were then students who quit high school to join the heroic atmosphere of the moment. We had, of course, many books in our knapsacks. The ones I brought with me were Ernest Hemingway and Jack London, two authors whose novels and short stories had been translated into Vietnamese and were prized by my parents," Khue said.

I have since learned that Hemingway's *A Farewell to Arms* remains required reading by Vietnam's general education program. What draws Vietnamese publishers, translators and readers to Hemingway and Jack London are the authors' representations of the poor and the struggle to survive in a harsh world. Before the advent of American Corners, the United States Department of State-sponsored initiative highlighting American cultural history and literature, the Vietnamese revealed how they obtained translated copies of Hemingway's works to better know us.

Huu Thinh, Vietnam's poet laureate and Chairman of the Vietnam Writers' Association, revealed that as a teenager in order to learn more about US culture and how Americans think he was led to read Hemingway's works. Thinh underscored that Vietnam was still a poor country but was quick to point out that Hemingway's stories have been taught in secondary schools for generations. He believes that Hemingway's stories and their expression of humanity and stories are in line with Vietnamese beliefs that favor kindness and raising voices against cruelty, injustice, and war.

Thinh says that *The Old Man and the Sea* is naturally linked to Vietnam since it is a coastal country and there is love for nature in the story. As a former tank commander and war correspondent, he was quick to acknowledge that during the war Vietnamese soldiers wanted to know the enemy and sought out translated editions of books like *For Whom the Bell Tolls, The Old Man and the Sea*, and Jack London's *Call of the Wild.*

There are few studies that focus on Vietnam's willingness to translate books and explore the culture of their enemy, but this was the case before the Communist Viet Minh forces defeated the French garrison at Dien Bien Phu. The poet Thinh asserts that the Vietnamese learned French and American literature very early, long before the wars took place, as these works reveal their enemy's civilization and humanity.

While Vietnam's approach to cultural diplomacy remains nascent, the Diplomatic Academy of Vietnam and the Academy of Journalism and Communication have established some cultural diplomacy strategy courses. A key pillar is building mutual understanding with other countries, especially the United States.

Anyone who visits Hanoi appreciates how passionate the Vietnamese can get about their soccer and literature. La Thanh Tung, a Vietnamese writer and contributor and former deputy editor at the prestigious literary and arts newspaper *Van Nghe*, spoke about the Hanoian admiration for culture and literature best reflected at their Temple of Literature.[8]

"It's most unfortunate that we know so much about American writers, like Hemingway, but that Americans know so little about our writers," Tung lamented. He also revealed that Bao Ninh, author of *The Sorrow of War*, was an unabashed fan of Hemingway.

There is now an even greater appreciation and reading of literature because of wider circulation of Vietnamese writing through English language publishing houses, and through greater access by the children of Vietnamese migrants to other countries like Canada and America.

It is also interesting to note how many younger Vietnamese poets have Beat poets or hip-hop music influences. These generational conversations will allow greater access to work and ideas honoring the Vietnamese understanding of the Cuban fisherman, Santiago, in *The Old Man and the Sea*. In conversations with Hanoi writers, their appreciation of the old fisherman is closely bound to the population's understanding of the hardships of all their fishers, men and women. These unsung heroes depend on the sea and weather for their living. They typically, like the old fisherman, get up halfway through the night since most fishing takes place before dawn. Some start working as early as 2 a.m. After they return from sea, they need to transport the fish caught to the market and clean and repair their fishing gear. Those on trawlers usually stay out at sea for three days and nights.

Still in possession of his well-worn translated copy of Hemingway's *The Old Man and the Sea*, Tung says, "I think the sea is large enough for fishermen to catch their fish but maybe they must venture out further than they should go."

Part III

Science Cooperation and Diplomacy

17

Marine Environmental Issues Call for Open Data and Science

Although Covid-19 has exacerbated global vulnerabilities, inequalities and fragilities, it may have also tapped possibilities for science and shared data cooperation among the ten nations that make up the Association of Southeast Asian Nations (ASEAN). Collaboration between Brunei, Cambodia, Indonesia, Laos, Malaysia, Myanmar, the Philippines, Singapore, Thailand, and Vietnam is not only curbing the health storm sweeping across borders, but also through ocean observatories that may succeed in reframing policies in the South China Sea.

Within ASEAN, there's a growing collective realization that sustainable development should be a central tenet in their marine conservation efforts and that it can be achieved by participatory community science efforts and digital monitoring tools. Their confidence is bolstered by China's pledge to make 2021 the China-ASEAN Year of Sustainable Development Cooperation and to step up cooperation on climate change, ecological conservation, environmental protection, and disaster mitigation.[1]

Just as the pandemic has made the world more dependent on digital platforms for communication and information, there's also been a developing tsunami of state-of-the art ocean observation systems, from Remote Operated Vehicles (ROVs), Autonomous Underwater Vehicles (AUVs) and towed sleds, wave gliders, sail drones, and large image collections. These new tools, including marine image informatics, have created a growing demand for non-professionals or citizen (community) scientists to screen the images and to communicate to the general public on ocean environmental threats.[2]

Credit must be given to Vietnam, the past chair of ASEAN. It has effectively used webinars to facilitate cooperation and has staged summits to tackle the pandemic, while also seeking solutions to many of the intractable problems in the South China Sea. In a disclaimer, I was an invited panelist at the 12th Annual South China Sea Conference, held November 16–17, 2020 in Hanoi and via webinar, on the subject of "Marine Science: how science and technology may impact on the good order of the sea."

With over 40 million observations of some 115,000 marine species from 1,600 datasets provided by nearly 500 institutions in 56 countries, the Ocean Biogeographic Information System houses the largest single data repository for biological data for the world's oceans. Given that national boundaries have little meaning with respect to environmental problems, there's an urgent need for establishing a participatory framework for achieving science co-operation and ensuring ecological security in the seas.[3]

"We need fundamental changes in the way that researchers work with decision makers to co-create knowledge that will address pressing development problems," claims Linwood Pendleton of the World Wildlife Fund.[4] In addressing ocean sustainability, he is quick to add that researchers need to share their data more freely and more quickly so that their work can inform decisions in real time and enhance the flow and exchange of information between communities.

A focused regional or ASEAN ocean stewardship initiative among scientists and the public offers pathways to address climate change, coral reef destruction, biodiversity loss, pollution (especially plastics), and fisheries depletion. It's clear that large amounts of ecological data are required, and that public participation can help analyze, collect, and categorize scientific data to assist in marine conservation and science cooperation. However, it's not enough that scientists have access to remote sensing platforms. The question is, how should marine scientists generate knowledge from all the data swirling around?

More importantly, are they capable of sharing the scientific knowledge with competing nation states operating over the same body of water? With emerging new data, there's an urgency for transparent and open access to science information and to place this uploaded data into a larger accessible digital ecosystem. The expansion of cabled observatories now brings data ashore through the internet. An open access of information sharing in the South

China Sea can benefit all claimant nations in the form of ocean governance and fisheries sustainability. The promise of open access awareness offers an opportunity to establish a regional marine science outreach for a possible Big Data South China Sea science community.

"Using big data and other scientific and technological means to facilitate Sustainable Development Goals has become a global consensus," says Zhang Yaping, vice president of the Chinese Academy of Sciences.

Ocean knowledge and technology are more developed today than ever before. Despite significant oceanographic advances and a continuous flow of ocean data, marine research has failed to ameliorate the competing South China Sea claims nor to navigate the sustainable stewardship of ocean resources. In this sea of opportunities, uncertainties, and threats, environmental degradation remains at the center of scientific conversation as marine scientists and citizen scientists sound the alarm about how to address issues of acidification, biodiversity loss, climate change, destruction of coral reefs, fishery collapses, and pollution—especially plastics. What's certain is that these ecological challenges reveal how claimant nations—the People's Republic of China, Vietnam, the Philippines, Malaysia, Brunei, and Taiwan—have a legal and ethical responsibility to ensure that none of their activities harm or create additional long-term damage to one of the most biologically diverse marine ecosystems.

The ocean, a natural laboratory, invites communities, governments, institutions, and scientists to understand the web of interconnections that links all of us. Fortunately, the emergence of marine citizen or community science stands at the intersection between ocean science and ocean literacy. Their contributions ensure that nations are poised to capitalize on the UN Decade of Ocean Science for Sustainable Development to create an open global data network, especially in the disputed sea. The timing is urgent to support efforts to reverse the cycle of decline in ocean health and gather ocean stakeholders worldwide behind a common framework that will ensure ocean science can fully support countries in creating improved conditions for sustainable development of the ocean.

Marine science has entered the digital age. Expansions in the scope and scale of ocean observations and smart sensors, now lead to a continuous flood of data.[5] This is an infrastructure for global, regional, and coastal sub-sea observatories now planned to support individual and networked sensors. These seafloor observatories enable real-time, continuous, and long-term observations that promise major breakthroughs in ocean sciences. The effort to dynamically control in situ sensor systems performing individual and cooperative observations tasks is both a challenge and a guarantee for the stable operations of functional observatories.

As a result, this provides opportunities to transform the way the ocean is studied and understood through more complex and interdisciplinary analyses and in coastal community engagement in the management and monitoring of marine resources.

"Big data sets are continuing to be developed at a global level. Competence and expertise remain very high in countries around the South China Sea. So, access to these should be no problem. Synoptic and in situ instrumental is increasing and there is greater gain with data sharing rather than data hogging," claims Dr. Liana Talaue-McManus, a biological oceanographer and who was previously an associate professor of Marine Affairs at the University of Miami Rosenstiel School of Marine and Atmospheric Science.

The improvement of global understanding of our oceans and their value will rely on innovation that removes barriers of access for the marine data needed among user nations. The ocean data networks will ensure cooperative monitoring for maritime safety and forecasting typhoons.

These collective science actions are essential because the focus of policy and research is directed toward understanding the critical changes that are occurring in the ocean systems. In conversations with fishermen, marine scientists, oceanographers, and student volunteers (citizen community scientists), they concur that there's a growing bandwidth of ocean data to draw upon.

During this pandemic, where science remains at the epicenter in the urgent production and distribution of a vaccine, there still exist impediments for international cooperation. Even so, ASEAN cannot always agree on or chart a course to promote sustainable ocean governance. And yet, sustainability and resilience become more important with the pandemic and its consequences make unprepared communities more vulnerable.

Mark Spalding, president of the Ocean Foundation, asserts that the 625 million people of the ten ASEAN nations depend upon a healthy global ocean. Meanwhile, coral reefs are dying as a result of an ecological catastrophe unfolding in the region's once fertile fishing grounds. As reclamation destroys more marine habitats, agricultural and industrial runoff poison coastal waters, and overfishing depletes fish stocks, it is no wonder that marine biologists are expanding their conversations and attempting to engage more citizens about the importance of using a rules-based ecological approach to protect the environment. Through studying the sustainability of the biological seascape, marine biologists are rallying for public access to ocean data that can be shared to respond to the damage done to the global commons.

As the world becomes more interdependent, global governance, including global economic governance and the governance of the global commons, is increasingly relevant for achieving sustainable development. For certain, the deepening economic globalization, and increasing migration, trade and capital flows, and climate change and increased activities in the global commons—those resource domains that do not fall within the jurisdiction of any one particular country, and to which all nations have access—make individual States more susceptible to policies adopted by others.

Enter Big Data and Marine Citizen Science

Although nation states take different approaches toward ocean data collection, all agree that data should be shared. In an e-mail from Dr. Peter Neill, director of the World Ocean Observatory, he revealed that the South China Sea is a case in point. He enumerates what we do with all the data and available technologies and their applications: first to our general compendium of knowledge; second, to our specific scientific interest; and third, to others who may use the data differently. Then he asks, in what form are the conclusions assembled and communicated beyond the narrow scientific audience to the public? He adds that "scientists and their sponsoring institutions are not well prepared to address the latter questions." Neill, like many other scientists, believes that all too often, science remains in the computers, labs, and ambition of scientists, and the public is denied access to the ocean data to better understand what must be protected, and understood.

Rick Bonney, a visiting scholar at Cornell University's Center for Engagement in Science and Nature, advocates for participatory research that enlists the public in collecting large quantities of data across an array of habitats and locations through the use of new technology, in the same way that smartphones and free downloadable science apps transform the collection of science data.[6]

Because of the increasing imperatives to predict changes, to monitor, and to protect coastal communities, marine scientists are reaching out and training public volunteers and citizen scientists in the collection and analysis of marine data, as a way to broaden public engagement. With photographic documentation and marine identification tools increasing, there's been a spate of free web and smartphone science apps and volunteering organizations, such as the Citizen Weather Observer Program, the Global Coral Reef Network, the Nature Mapping Foundation, and the Marine Debris Tracker. Available to the public, these digital tools effectively enhance scientific literacy, deepen connections to nature and place, and foster new knowledge networks. These tools offer so much more daily promise as more passionate citizen scientists from all over the world record data with easy-to-use apps, contributing to an open data platform.

Vietnam has encouraged remote sensing and citizen science to fill in the gaps of conventional environmental monitoring methods. In the past, Vietnam's scientists examined national water monitoring infrastructures but failed to fully account for the information received from free satellite images and crowd-based data collection.

The links between citizen or community science and the power of ocean data revolution are clear, and the correlation between marine technologies and the pooling of data sets helps influence policy development. This engagement with citizen scientists extends from California to Vietnam's Cu Lao Cham where young people are helping with marine protected area monitoring.

According to citizen science researchers Cathy Conrad and Krista G. Hilchey, "there's a wealth of community-based management initiatives around the globe." In their academic work, they reinforce that it's not just the traditional role as "scientists using citizen science as data collectors, but rather citizens as scientists."[7] In the South China Sea, there is increasing need for tools for monitoring especially when fragile ecosystems are declining from the desired state.

The coral reef at Dongsha Atoll, near southeastern China and the Philippines, is now dead. It was once rich in marine life and regarded as one of the world's most important coral reefs. Marine scientists have repeatedly warned of the devastating coral loss due to a spike in sea temperatures. This was especially true in 2016 where coral bleaching was the worst on record for Australia's iconic Great Barrier Reef.

Furthermore, some of the South China Sea reefs are now dead due to China's dredging and developing military bases atop them. At an East West Center webinar last September entitled "A Crucible for Turning the Tide in the South China Sea," Professor John McManus emphasized that "the coral reefs of the South China Sea are among the most diverse in the world. For instance, they have about 600 species of coral in the Philippines and 571 in this area east of Palawan." However, McManus made clear that the overwhelming majority of Chinese construction that occurred between 2014 and 2017 resulted in severe damage to coral reefs found at Fiery Cross Reef, Subi Reef, Mischief Reef, Woody Island, and some other islets.

By his estimates, roughly 100 square miles of reefs have been destroyed by base-building and clam hunting. "The bases have irreversible damage," he says. "A square kilometer is a million square meters. 15 square kilometers of permanent damage that can never be replaced and 90 percent of that is from the bases China was building and dredging even harbors and channels. The damage is simply widespread across many islands."

McManus is not the sole voice calling for effective coordination and monitoring of the ocean's coral reefs. He believes that best mechanism for this coordination is a South China Sea regional fisheries organization. He was joined in our panel by Dr. Nguyen Chu Hoi, a former deputy administrator at Vietnam's Administration for Seas and Islands and an associate professor of Marine Science and Governance Policy at Vietnam National University in Hanoi.

Chu Hoi agrees that Chinese reclamation not only destroys the coral reef formations, but also the connectivity of some of the marine species. "We need healthy reefs to provide food and fishery resources not only for Vietnam but for all of the countries dependent on the South China Sea."

Also, Dr. Ma Carmen Ablan Lagman, professor at De La Salle University in the Philippines and fellow panelist, also weighed in on the marine environment scene in the South China Sea. She has also been witnessing this critical level of marine degradation and the depletion of fish stocks in the contested waters. The political and economic dynamics of regional countries continue to generate adverse consequences for the marine environment. She contends that the UNCLOS may be the single most important regional policy instrument that has changed the face of fisheries in the region and in the world. "With this policy," she said, "we actually have the ability to call a certain part of the ocean our own, which brings with it the ability to [invest more], because of knowing that we will have a return on those investments."

In her presentation, Lagman presented a macro picture of the fishery industry in the region. She identified countries like China, Indonesia, the Philippines, Malaysia, Thailand, Myanmar, and Taiwan that are possessing the highest marine capture fisheries production. Therefore, fish plays a huge part in the diet and importance of livelihoods for millions in the region.

Although the pandemic is regarded as one of the world's greatest challenges, Earth Challenge 2020 was initiated to gain the global support for public volunteers or citizen scientists to become the world's largest coordinated citizen science campaign for data collection, and as a platform for global citizen science data. It does take a village to monitor coastal fisheries and to protect marine environments.

Through mobile apps, people around the world are able to monitor threats to environmental health in their own communities. This new open data platform is making it easier for researchers around the world to find and access high-quality information for international policy assessments like the UN Sustainable Development Goals.

In Australia, a coalition of students, environmental groups, universities, and scientists are gathering critical new data about microplastics in the ocean and their waterways. The data collected by a network of citizen scientists and researchers enables AUSMAP to create vivid maps of microplastic hotspots in the country. According to research scientist, Dr. Michelle Blewitt, "our work enables communities and government to implement behavior change, regulate industry and develop better waste management."[8]

Civil society activities related to South China Sea dispute management are politically limited and not widely accessible to the public. However, Vietnamese environmental organizations like the Center for Development of Community Initiative and Environment, Mekong Environment Forum, Mekong Delta Youth, Marine Life Conservation and Community Development, have missions to solve environmental issues in the South China Sea. They have been educating all of society, especially young people, fisheries, and businessmen. In that sense, environmental advocacy can translate into successful diplomatic efforts and the democratization of science.

New Marine Technologies Connect Data but not Always Nations

While ocean technology advances have multiplied over the past decade, including the scale and number of cabled observatories, acoustic modems, and processing and visualization capabilities, cooperation among nations to foster an open access digital ecosystem requires more development. According to the Blue Paper report "Technology, Data and New Models for Sustainably Managing Ocean Resources," vast stores of ocean data remain restricted in the databases of governments, researchers, and industry.[9]

Technology has ushered in a variety of ocean observing systems and new available data sources that are enabling a wide array of digital tools and thus information flows to policy experts, marine resource personnel, and citizens. The North Pacific Marine Science Organization, also known as PICES, is an intergovernmental scientific organization promoting and coordinating marine research in the North Pacific and adjacent seas. It was established in 1992 by China, the United States, Canada, Japan, Republic of Korea, and Russia. Their goal is to advance and to collect scientific knowledge about the ocean environment, global weather, climate change, and marine ecosystems.

Dr. Sara Tjossem, a senior lecturer in International and Public Affairs at Columbia University, claims that PICES from the outset "struggled with how best to exchange data, but not become a redundant data repository." In Tjossem's research, she calls for all "data to be quickly accessible to reflect real world events and readily exchanged physical data proved easier to share while chemical and biological data were more challenging."[10]

It is promising that an increasing number of marine scientists recognizes that the South China Sea is a natural laboratory for science collaboration. The mantra is global and simple: there should be no national borders in science. The focus is to rise above politics and seek solutions on the larger and important question central to humanity's long-term wellbeing.

While it may be too much to hope that South China Sea nationalism will disappear because of advances in technology or community-wide science participation to ease cross-border environmental issues, the creation of a South China Sea open-access domain awareness model may inspire public good from governments, NGOs, and fishers. All of this will require proofs of concept. But there's much to gain and too much to lose if nations don't try.

18

Managing the South China Sea Commons through Science Policy[1]

The South China Sea is the home, for now, to some of the most spectacular biodiverse coral reefs in the world. Yet the world continues to witness satellite images of these troubled waters that show the rapid destruction of such extraordinary reefs. The cause of this ongoing destruction, which amounts to nothing less than a widening environmental crime scene, is the reckless land reclamation activities conducted by the People's Republic of China as it attempts to turn rocks into islands and to bolster its expansive claims.

In this sea of opportunities, uncertainties, and threats, environmental degradation remains at the center of scientific conversation as an increasing number of marine scientists sounds the alarm on issues of acidification, loss of biodiversity, climate change, destruction of coral reefs, and fishery collapse.

IUU fishing has clearly emerged as a key maritime threat to Asian seas, including the South China Sea. With annual catch production accounting for over 10 percent of global total, the fishery resources are very important to the coastal areas of the region, where over 77 percent of the population rely on the pelagic fishery resources for their daily protein intake and livelihood.[2] Increasingly, observers and fishery experts acknowledge that excessive fishing is the primary threat to future fisheries sustainability in the SCS. Unfortunately, this condition has continued in both commercial and small-scale fisheries.[3]

There's much agreement that the lack of strong fisheries governance has perpetuated this threat, which in turn spurs other problems such as a rise in IUU activities.[4]

With environmental security shaping a new South China Sea narrative about ecological challenges, this concept represents a crucial effort to link the impact of environmental change to both national and international security. Paul Berkman, oceanographer and former head of the Arctic Ocean Geopolitics Program at the Scott Polar Research Institute and founding director of the Science Diplomacy Center at the Fletcher School of Law and Diplomacy at Tufts University, provided his definition of environmental security. "It's an integrated approach for assessing and responding to the risks as well as the opportunities generated by an environmental state-change," he claims.

On September 30, 2020, I organized a webinar program, "Science Diplomacy: A Crucible for Turning the Tide in the South China Sea, in cooperation with the Washington-based East West Center, Professor Berkman gave a key note address on the application of science diplomacy in the region. He believes that there are a series of challenges in building a conservation consensus among nations.

"There is the challenge of sustainability time-scales. This includes the elements of the natural ecosystem, the species in the system, as well as the geopolitics of the region. All of that is part of the deriving balance. And the challenge in this is to think beyond the moment—beyond the political, geopolitical context of super powers, allies, and adversaries. The challenge is to extend the dialogue and consider urgencies that operate across generations and to look into the future in a way where the questions themselves can facilitate common interest building."

The Spratly Islands are the focal point of a territorial dispute that represents a serious threat to regional security in Southeast Asia. Six governments—China, Vietnam, the Philippines, Malaysia, Brunei, and the Republic of China (ROC) on Taiwan—have all laid claim to all or some of the more than 230 islets, reefs, and shoals in the Spratly area.

However, the unanimous decision reached in the summer of 2016 by the Hague's tribunal found that China's large-scale reclamation and construction of artificial islands has caused severe harm to coral and violated the country's obligation to preserve fragile marine environments. Furthermore, it denied China's legal basis to claim historical rights over a vast majority of the South China Sea. It was a striking victory for the Philippines, which led the case.

Among many dramatic findings, the tribunal declared China's so-called "nine-dash line" invalid.

"The Tribunal has no doubt that China's artificial island-building activities on the seven reefs in the Spratly Islands have caused devastating and long-lasting damage to the marine environment," reads the judgment.

In addition, the UNCLOS stipulates in two out of 17 parts a direct application to the merits of marine science research with an emphasis on encouraging bilateral and multilateral agreements to create favorable conditions for marine science study.

Destructive Practices

Professor John McManus, marine biologist at the University of Miami and notable coral reef specialist, who has regularly visited the region and provided analysis to the tribunal, has stated that based on satellite imagery the environmental damage done by Chinese dredging and clam poaching is most severe.

McManus has researched this region for more than a quarter of a century. He knows that the most important resource in these heavily shed waters is the larvae of fish and invertebrates. As a result, he has called repeatedly for the development of an international peace park in this contested region.

"Territorial disputes have led to the establishment of environmentally destructive, socially and economically costly military outposts on many of the islands. Given the rapid proliferation of international peace parks around the world, it is time to take positive steps toward the establishment of a Spratly Islands Marine Peace Park," said McManus.

Because this region is one of the most important large marine ecosystems in the world, rich in marine living resources and a mainstay for livelihoods of local communities in South China Sea nations, policy shapers are taking note of the necessity to create more marine protected areas.

"Vietnam, a claimant nation, already has marine reserves, so this involves extending this practice to the Spratly area. It is usually to institute conservative harvest and protection practices when there is the threat of competition from outsiders. Vietnam stands to lose a great deal if the current situation continues and results in a general decline in fisheries across the South China Sea," claims McManus.[5]

Policymakers may do well to take a lesson or two from nature as they examine how best to address the complex and myriad sovereignty claims.

The marriage of policy and science is essential to navigating these perilous geopolitical waters. The concept of science diplomacy is not a new paradigm, but it embraces collaboration and adroitly addresses problems related to environmental protection where they arise.

In the East West webinar, McManus presented scientific evidence about the environmental wreckage associated with China's reclamation projects that involved dredging to create military installation in seven locations in the disputed sea. "The coral reefs suffered irreversible damage that can never be replaced and 90 percent of that is from Chinese dredging harbors and channels and their use of giant clam chopper boats. Now we thought that China had stopped this in 2016 but they kept going and continued in Scarborough Reef as well as the Paracels. This is a problem. However, if they stopped now, the reef will recover mostly within twenty years," says McManus.[6]

The passionate ecologist's urgent call to set aside sovereignty claims to address the protection of coral reefs and overfishing is embraced by more ASEAN states, and surprisingly among some Chinese policy experts. In the face of deteriorating marine ecological environment, Chinese government and authorities of coastal areas have adopted a series of countermeasures, including developing marine protected areas.

Over the past decade, Chinese officials are understanding the need to establish marine protected areas. "By the end of 2017, 270 marine protected areas have been established in China's coastal areas, representing a total area of 120,000 square kilometers accounting for 4.1 percent of the total sea area. There are 35 national marine sanctuaries representing a total area of 17,751 square kilometers and 67 special marine reserves, with a total area of 7,250 square kilometers," says Wang Bin from the State Oceanic Administration.[7]

Furthermore, in a recent CSIS webinar held on January 13, 2021, Taiwan's Foreign Minister, Joseph Wu acknowledged their nation's efforts to utilize their Taiping Island (Itu Aba), one of the largest in the South China Sea, as a station for international science research and cooperation.[8] To be clear, the island although administered by the ROC, it is also claimed by the PRC, the Philippines, and Vietnam. In 2016, in the ruling by an arbitral tribunal in the Permanent Court of Arbitration, in the case brought by the Philippines against China, the tribunal classified Itu Aba as a "rock" under UNCLOS and therefore not entitled to a 200-nautical-mile EEZ and continental shelf.

"We will continue doing humanitarian search and rescue work in the South China Sea and we view our role as a peacemaker and as a central location for marine science cooperation," says Minister Wu. In many instances,

Taiwan's naval personnel have also been engaged in assisting both Chinese and Vietnamese boats in distress during the annual stormy typhoon season. Even former President Ma Ying-jeou had previously called upon other claimant nations to share resources.[9]

The ASEAN Working Group on Coastal and Marine Environment recognizes that the region faces enormous challenges to sustainability in coastal and shared ocean regions. Unless a scientific approach to the ecosystem is adopted, transboundary conflicts in marine areas can and will get worse.

In an *Asia Times* op-ed that I co-authored with Leonard Hammer, we wrote how Taiwan's Taiping Island is an ideal location for a global science peace park. Appropriately, the island's Chinese name, Taiping, translates as peaceful, and as such affords a sanctuary for a common interest approach to the region's acrimonious disputes.[10]

Adjusting for the reality that Taiwan actually does control and hold Taiping Island opens the door for more effective and beneficial use of the island. That is, Taiwan has continuously asserted sovereignty over the area, thus cementing its control over the island, and since 2008 also has begun to propose the idea of turning the area into an internationally protected environmental research zone. This action is borne out of the reality of climate change. All evidence suggests that the island with its many seabirds, sparrow hawks, papaya plants and assorted coconut trees as well as a lighthouse, renovated airstrip and hospital, will eventually become submerged, or at the very least, subject to extreme weather conditions that will likely get worse, making it costly and non-beneficial for Taiwan to maintain the island.[11]

Since ASEAN's inception, it has been occupied with the task of identifying shared solutions to common security problems. To a large degree, one may say that security questions have been the driving force for continued regional integration in Southeast Asia. In the future, questions of environmental security may play the same role.

According to Karin Dokken, a political scientist at the University of Oslo, "The states around the South China Sea are to a large degree interdependent when it comes to questions of the human environment.

"They are interdependent to the degree that if they fail to end common solutions to environmental problems they may end up in violent conflict against each other." Dokken said, "In general, environmental interdependence is both a source of conflict and a potential for international integration."[12]

Without agreement on these environmental problems, there is a bleak future for the sea. Nearly 80 percent of SCS coral reefs have been degraded and are under serious threat from sediment, overfishing, destructive fishing practices, pollution, and climate change. There are additional disturbing marine pressures associated with the environmental degradation since the Spratly Islands constitute a distinct marine ecosystem, which serves as a significant source of larvae for regional coral reefs. In fact, the biological impact of the dredging operations, especially at Mischief Reef, are being studied.[13]

Challenges around food security and renewable fish resources are quickly becoming a hardscrabble reality for more than just fishermen. With dwindling fisheries in the region's coastal areas, fishing state subsidies, overlapping EEZ claims, and mega-commercial fishing trawlers competing in a multi-billion-dollar industry, are at the core in this sea of troubles.

An ecological catastrophe is unfolding in the SCS's once fertile fishing grounds, as repeated reclamations destroy reefs, agricultural and industrial runoff poison coastal waters, and overfishing depletes fish stocks. This dire situation is compounded with climate change. Scientists project that if the warming continues at the present trajectory, record-breaking heat waves would be up to 20 times as likely toward the end of the 21st century.

In 2014, the Center for Biological Diversity warned that it could be a scary future indeed, with as many as 30 to 50 percent of all species possibly headed to-ward extinction by mid-century. Fish catches have remained at an unsustainable 10–12 million tons per year for decades—a number that could double when IUU fishing practices are included.

After all, the United Nations Environmental Program confirms that the South China Sea accounts for as much as one tenth of global fish catches, and by 2030 China will account for 38 percent of global fish consumption. Overfishing and widespread destruction of coral reefs now necessitate the intervention of science policy to safeguard stewardship of this vital area.

China has been at the forefront of this major fish exploitation. With over 2,000 blue water commercial trawlers and over 100,000 fishing vessels, including a 3,000-ton processing ship, the evidence seems compelling that Beijing is not only responsible for the destruction of coral reefs but it is also contributing to fishery collapse.

Foreign Policy magazine asserts that these fishing incidents and direct acts of violence are significant "because it underscores how central fishing is to the simmering territorial disputes that are turning the South China Sea into a potential global flashpoint—and how far countries are willing to go to defend their turf, or at least what they claim is theirs."

Fishing remains a politically sensitive and emotionally charged national security issue for all claimant nations, but ocean plundering presents the region with a looming food crisis. Any effort to balance the economic benefits with the security context within the South China Sea will require a coordinated, multi-level response from scientists, who have historically engaged in collaborative research and already been addressing issues of sustained productivity and environmental security in the region.

The immense biodiversity that exists in the South China Sea cannot be ignored. The impact of continuous coastal development, escalating reclamation and increased maritime traffic is now regularly placed in front of an increasing number of marine scientists and policy strategists.

Marine biologists, who share a common language that cuts across political, economic, and social differences, recognize that the structure of a coral reef is strewn with the detritus of perpetual conflict and represents one of nature's cruelest battlefields.

While traditional diplomatic and military tactics are not completely exhausted in the latest rounds of diplomatic salvos between China and the United States, perhaps the timing is excellent for the emergence of science as an optimal tool for bringing together various claimants, including Brunei, China, Malaysia, the Philippines, Vietnam, and Taiwan in the highly nationalistic, contested sea disputes.

A joint scientific declaration for urgent action on an environmental moratorium on dredging is much needed. Recent biological surveys in the region and even off the coast of China reveal that the loss of living coral reefs presents a grim picture of decline, degradation, and destruction. More specifically, reef fish species in the region have declined precipitously, from 460 species to around 261.[14]

After all, this environmental change is a global issue that holds no regard to sovereignty. The destruction and depletion of marine resources in the Spratly Islands harms all claimant nations. Perhaps citizens from the region who are directly impacted by the environmental attack on their sea and their fragile coral formation can connect with a coral reef action network.

In fact, there's an existing innovative international organization, the International Coral Reef Action Network (ICRAN), established in 2000 and composed of many of the leading coral reef and conservation organizations. ICRAN consolidates technical and scientific expertise in reef monitoring and management to create strategically linked actions across local, national and global levels. They are the first alliance to respond to conservation needs at a global scale by recognizing both traditional and scientific perspectives of coral reef dynamics and respective social dependency. Their actions to build resource stewardship within communities helps provide opportunities to develop the skills and tools needed to ensure the sustainable use, and long-term vitality of coral reefs.

Protected marine reserves are an emerging tool for marine conservation and management. Sometimes called "ecological reserves" or "no-take areas," these marine protected areas are designated to enhance conservation of marine resources.

Vietnam, a claimant nation, is wasting little time responding to the region's environmental challenges and is fast-tracking its own model marine protected area program. Cu Lao Cham is located about 20 kilometers off of Vietnam's central coast.

The Cham Islands are a marine protected area (MPA) that was established by the Provincial People's Committee of Quang Nam Province in December 2005. Professor Chu Manh Trinh, a 57-year-old Da Nang University biology professor, is responsible for mapping out the agreed-upon objectives of protecting the natural resources and the cultural and historical values of the Cham archipelago. In 2009, the area was designated a World Biosphere Reserve by UNESCO.

Vietnam has adopted MPAs to address present and future food security issues. These MPAs play an important role in the development of the marine economy; they improve livelihoods in coastal fishing communities, and also serve to protect national sovereignty claims.

The region needs to bring together the most qualified scientists who have experience studying marine biodiversity and environmental sustainability in the troubled SCS waters to participate in science policy forums. Their collaborative work may lead to the successful development of a South China Sea. As a result, their scientific efforts may then inspire ASEAN to further cooperate in responding to regional resource management by issuing a call for a moratorium on any further damaging reclamation work.

Furthermore, UNCLOS, to which all claimants in the SCS are signatory parties, is clear on the need for international cooperation in the resource management of seas such as the SCS. Article 123 states that: "States bordering an enclosed or semi-enclosed sea should cooperate with each other in the exercise of their rights and in the performance of their duties under this Convention. To this end they shall endeavor directly or through an appropriate regional organization follow these requirements: to coordinate the management, conservation, exploration and exploitation of the living resources of the sea; to coordinate the implementation of their rights and duties with respect to the protection and preservation of the marine environment; to coordinate their scientific research policies and undertake where appropriate joint programs of scientific research; to invite, as appropriate, other interested organizations to cooperate with them in furtherance of the provisions of this article."[15]

Of course, China has many excellent coral reef scientists of its own, and recognizes that it is in the best interest of Beijing to protect coral reefs, maintain sustainable fisheries, and to eventually avail themselves to eco-friendly tourism once tensions decline.

The common ground shared by all claimants is that an increasing number of South China Sea fisheries are hurtling toward collapse and this translate into a looming environmental security issue, and the outcome is all too likely to be conflict. The global scientific, conservation, and legal communities must unite to halt the coral reef destruction, biodiversity loss, and fisheries depletion.

The tribunal's first international ruling on the South China Sea offers an opportunity for measured steps toward peace and security. Of course, ASEAN has demonstrated a weak institutional capacity to address complex political and environmental issues, but the rest of the world, including the United Nations and Washington, is watching carefully how international law and its application on various claims can lead to a peaceful and lawful path forward.

Actions Needed

Claimant nations' marine scientists and policy shapers might take up some of these confidence- building options:

- Establish complete freedom of scientific investigation in the contested atolls and reclaimed islands.
- Expand science cooperation among all ASEAN marine scientists through more academic workshops.
- Place aside all territorial claims.
- Create a regional Marine Science Council to address environmental degradation issues.
- Foster dialogue for a proposed marine peace park.

- Appoint a science-led ASEAN committee to study the Antarctica Treaty and the United Nations Environmental Program initiative under the East Asian Seas Action Plan.
- Propose a renaming of the contested sea to the Freedom Sea or the Southeast Asian Sea.

If there are to be any fish left in the contested sea, an ASEAN ecological agreement—led by Brunei, Malaysia, the Philippines, and Vietnam—can steer others to unite around a proposed international peace park or, at the very least, a cooperative marine protected area situated prominently in the Spratlys.

It's the first step in supporting trust and confidence among neighbors and in implementing a common conservation policy. After all, coral reefs are the cathedrals of the South China Sea. It is time for more citizens and policy shapers to rally around marine scientists so that they can net regional cooperation and ocean stewardship to benefit all before it is too late.

19

Ecological Politics and Science Policy in the South China Sea[1]

Rival countries have wrangled in the South China Sea over a string of atolls, coral reefs, and islets for centuries, but now these competing claims are viewed as a serious challenge to peace and prosperity in the region. These disputes that are associated with continuous coastal development, escalating reclamation, and increased maritime traffic in the sea, also draw attention to the destruction of coral reefs and the overall environmental degradation in the troubled waters, and reveals how claimant nations—China, Vietnam, the Philippines, Taiwan, Malaysia, and Brunei—have a legal and ethical responsibility to ensure that none of their activities create harm or long-term damage to the fragile marine ecosystems.

Efforts toward diplomatic or legal solutions for this maritime flashpoint seem to have deadlocked all parties, including the ten-member ASEAN; since they cannot agree nor chart a course to promote sustainable governance without incurring economic loss or conceding sovereignty ground. Meanwhile, coral reefs are dying as a result of an ecological catastrophe unfolding in the region's once fertile fishing grounds. As reclamations destroy more marine habitats, agricultural and industrial run-off poison coastal waters, and overfishing depletes fish stocks, it's no wonder that more marine biologists are offering their voices and science to explain the importance of the environment in a rules-based ecological approach. Their efforts to respond to the damage done to the global commons through studying sustainability of the biological sea-scape as well as navigating the development of science diplomacy to prevent geopolitical battles over marine resources management will require scientific forums and collaborative problem solving among all neighbors.

Enter science diplomacy. Science has been adopted as a diplomatic tool for peace-building for several decades by many countries, including the United States. John F. Kennedy in 1961 established a science and technology coop-eration agreement with Japan in an effort to restore the intellectual dialogue between the two countries after World War II. Furthermore, there are many organizations that strengthen global scientific relationships. For example, the American Association for the Advancement of Science, formed in 1848, is the largest scientific organization in the world and reaches scientists world-wide. They also have a Center for Science Diplomacy that effectively builds scientific cooperation and collaborative bridges. The Center's journal, *Science and Diplomacy*, provides a forum for open policy discussion. Also, the Interna-tional Institute for Applied Systems Analysis (IIASA) was established in 1972 by representatives of the Soviet Union, United States, and 10 other countries from the Eastern and Western blocs as an effort to "use scientific cooperation to build bridges across the Cold War divide and to confront growing global problems on an international scale."[2] Since then, IIASA has been developing largely with 24 National Member Organizations with the mission to "bring together a wide range of scientific skills to provide science-based insights into critical policy issues in international and national debates on global change, with three central research focuses."[3]

Although nation states have different approaches regarding to science diplomacy, in general, this type of diplomacy can be identified with the main dimensions from the definition provided by the American Association for the Advancement of Science (AAAS): science in diplomacy (science to inform foreign policy decisions), diplomacy for science (promotion of international scientific collaborations), and science for diplomacy (establishment of scientific cooperation to ease tensions between nations).[4] In that sense, there is a widely accepted agreement among environmental policy planners that science diplo-macy can positively contribute to conflict resolution.

Peter D. Gluckman, a chief science advisor to the Prime Minister of New Zealand and chair of the International Network for Government Science Advice, promulgates how science influences policy. "But as we have seen in the complex processes associated with the Intergovernmental Panel on Climate Change (IPCC), sometimes very elaborate processes are needed for globally driven science to influence domestic policies and to diminish the role of national interests in shaping the science. It could be argued that the elaborate nature of the IPCC exercise was the inevitable outcome of a situation where very distinct national interests and values were at play with regard to the economics of climate change. Inclusiveness builds trust, so it was important for the IPCC to broaden the scope of expertise and clearly demonstrate that the scientific consensus was international."[5]

As such, science diplomacy is not a completely new approach to international relations, and in South China Sea dispute management in particular. However, at this moment, it seems that this type of diplomacy has raised two important questions in efforts to successfully settle the South China Sea dispute: Should we do it? And can it be successful? The chapter addresses both questions by examining the characteristics of science diplomacy in the South China Sea dispute from both a historical view and from analysis derived from scholars and policy shapers who are explicitly interviewed for their ideas on this issue.

Science Diplomacy—More Gains than Losses

If any cost-benefit analyses are applied to the methods used for settling the South China dispute, science diplomacy would be among the first ones selected because it can bring about many benefits and its implementation does minimal harm to the settlement process. As Paul Berkman, a recognized oceanographer and former head of the Arctic Ocean Geopolitics Program at the Scott Polar Research Institute, says, "Science contributes fundamentally to the implementation of sustainable development strategies that seek to balance environmental protection, economic prosperity, and social justice into the future."[6] Berkman's insightful understanding of historical and scientific perspectives in the context of environmental policies, both in the Arctic and in Antarctica, offers valuable lessons for possible adoption in the South China Sea.

At first, science diplomacy helps promote confidence building among the parties directly and indirectly involved in the South China Sea dispute. Science diplomacy characterized by scientific cooperation activities has contributed to solving many transboundary issues among nations sharing the same marine waters and in marine areas beyond national jurisdiction.[7] Environmental monitoring successfully offers a context for countries to express their true perception of the region without being affected by nationalistic, political, or economic factors like sovereignty or foreign policy objectives. As a result, it provides claimants and other involved parties in the South China Sea with an effective way to evaluate the political willingness of the partners and policy-makers as well as to better understand the overall picture in the South China Sea.[8] Consequently, the claimants can be more confident in future cooperation on other issues. In other words, science diplomacy can establish a useful and convenient starting point for regional cooperation to deal with not only the international environment problems but also the achievement of South China Sea settlement in particular and the region's prosperity and peace.[9]

Secondly, science diplomacy offers a much-needed strategic pause in the rising tensions in the South China Sea. The Hague's five judge tribunal ruling in July 2016 found that China's large-scale reclamation and construction of artificial islands has caused severe harm to coral reefs and violated the country's obligation to preserve fragile marine environments. Furthermore, it denied them any legal basis to claim historical rights over a vast majority of the South China Sea, and declared the nine-dash line as invalid. The tribunal ruled in favor of the Philippines, which filed the case. The situation in the roiling sea became highly sensitive when China was under domestic pressure to save face,[10] while the Philippines was met with challenges of having to "do something" after the legal victory.[11] More importantly, the competition in the South China Sea is not solely a resource or legal oriented one, but it's also a strategic and territorial dispute. As a result, any country's response without careful consideration risks triggering an armed conflict. Therefore, it would be wise for parties to pause and look for options to settle the dispute instead of desperately seeking some endgame solution. Science diplomacy can fulfill these requirements. In fact, cooperative science activities do not have any effect on the status quo of the South China Sea dispute. However, it keeps alive the hope for a solution to this dispute by creating a myriad of activities with all of the involved parties engaged especially in environment, economic, and security issues, instead of perpetuating the political stalemate.

The role of science diplomacy in solving illegal fishing in the South China Sea can be seen as an example. The illegal fishers have been used as sentinels in maritime territorial disputes, in which nations are already using naval forces to bolster their sovereignty claims. In the contested waters, clashes between the claimant governments and foreign illegal fishers continue. In that sense, the prospect of claimants in the South China Sea going to war over access to fishing waters is a real and immediate threat.[12] However, compared to issues like the sovereignty claims, the approach of science diplomacy to fishery collapse may be one of the most urgent but the least politically sensitive, one which can be solved without provoking the nationalism and other traditional concerns which are currently heightened in the region. In other words, science diplomacy provides the involved parties in the South China Sea dispute with a rational and transparent way to avoid the worst while looking for the best.

Finally, the benefits of science diplomacy can extend beyond the South China Sea disputes. While most of the political and military efforts toward the South China Sea disputes are limited to dispute conflict resolution or management, science diplomacy provides other collateral benefits. The claimant states have dealt with a wide range of non-traditional security issues through science diplomacy, tackling regional issues like transboundary crimes, economic development and environment protection, climate change, coastal pollution, coral reef destruction, overfishing exploitation, ocean acidification, and marine protected areas. In a recent article in *Philippine Daily Inquirer*, Fidel V. Ramos, the former president of the Philippines (1992–1998) and a member of the ASEAN Eminent Persons Group, stated that environmental cooperation can promise to bring about the "mutually beneficial efforts to improve tourism and encourage trade and investment, and to promote exchanges among think tanks and academic institutions on relevant issues."[13]

For example, the coral reef ecosystem services value in the South China Sea is $350,000 per hectare per year, which can be a valuable contribution to poverty alleviation, crucial to economic development in the region.[14] Another example is found in the project Reversing Environmental Degradation Trends in the South China Sea and Gulf of Thailand. It was established by the United Nations Environment Program (UNEP) and recognized as one of the very first acts of science diplomacy to deal with the South China Sea disputes at the regional level. Four out of the five main outcomes of the project went beyond the South China Sea dispute. They include "a series of national and regional management plans for specific habitats and issues; a suite of demonstration management activities at sites of regional and global significance; a regional framework system of fisheries refugia in the South China Sea and Gulf of Thailand; and pilot activities relating to alternative remedial actions to address priority transboundary pollutants and adopted water quality objectives and standards."[15] Over the next several years, rising demand from growing populations and economies, are bound of a direct collision course with exploitation, pollution, habitat destruction, climate change, and devastation.[16] A policy goal of mere sustainability seems inadequate;[17] science diplomacy practices in the South China Sea must now urgently address not only the disputes but also other aspects of international discord.

With the benefits science diplomacy can bring about, it should be placed in serious consideration among policy makers while finding an initiative to settle the South China Sea dispute.

Science Diplomacy—A Mission Possible

In order to be an effective science policy, one initiative needs to be not only beneficial but also feasible. And science diplomacy has proven itself to be an eligible one.

First, all claimant nations and other involved parties in the South China Sea disputes, to some extent, are now capable of adopting science diplomacy.

Comparing the expenditures that a country may use for other methods of solving the South China Sea disputes—from military to multilateral diplomacy or a summit—science diplomacy seems to be an affordable way for developing countries such as Vietnam or the Philippines. In fact, it is hard to draw an exact comparison of the expenditures a government provides for other ways of solving South China Sea disputes, but it is apparent that science diplomacy is among the few ways that even non-state actors can also join in and donate without any barrier. This is in sharp contrast to other situations where state agencies are the only actors authorized to fund these activities.[18] The reason for this difference comes from the way types of diplomacy are interpreted. While military and economic initiatives can be seen as the actions of one country protecting its sovereignty and population that is directly related to national security, a sensitive subject that any non-state actor involvement is inevitably considered inappropriate. The science initiatives can be more widely accepted as efforts to solve global issues that require contributions of all actors in the international community. For example, the 2009 UNEP project Reversing Environmental Degradation Trends in the South China Sea and Gulf of Thailand,[19] was joined by Cambodia, China, Indonesia, Malaysia, the Philippines, Thailand, and Vietnam. 53 percent of its funding came from participant governments, while 44 percent came from the Global Environment Facility Trust Fund, and an additional 2 percent from the United Nations Environment Facility.[20] This can be seen as a win-win solution. On the one hand, it makes science diplomacy-related initiatives financially possible, and on the other hand, it makes public policy effective with the aid of NGOs who are very good at sharing information, since they raise awareness, and build capacity at the community level. It also helps non-state actors to get what they want, namely new markets and economic opportunities, especially for those in the private sector that combine their business interests with corporate social responsibility goals, and it helps NGOs achieve the goal of comprehensively enhancing people's quality of life. In other words, economically speaking, science diplomacy is seen as all "carrot" with no "stick" for dispute management in the South China Sea.[21]

In political terms, science diplomacy is a safe approach for every government. In this regard, Dr. Sophie Boisseau du Rocher, a senior researcher and associate at the Paris-based Centre Asie at IFRI, claims that, "Scientific programs obviously serve the interest of both China and the Southeast Asian countries. Areas of cooperation are numerous and [these programs are] the easiest ground to establish regional cooperation regimes. In this sense, collective scientific work or the adoption of functional standards could contribute toward defusing threatening attitudes."[22] While economic or military cooperation requires a lot of foreign policy and national security consideration, cooperation in science is much more neutral, even in countries with a lot of conflicts, since they can cooperate with each other in scientific projects "to affirm and to improve human life" without worrying about their messaging to the international community about their foreign policy orientation or invoking domestic anger because of shaking hands with the wrong partners. This can clearly be seen in science cooperation agreements between political opponents like the US, the USSR, and China in the 1970s and 1980s in the middle of the Cold War,[23] or the Joint Oceanographic and Marine Scientific Research Expedition in the South China Sea (JOMSRE-SCS) that will include the participation of ASEAN member countries, China, and international organizations in the next phase.[24]

In social terms, science diplomacy can receive support from society more easily than any other type of diplomacy applied in solving South China Sea issues. It serves the essential needs of the human lives. In fact, while other types of diplomacy often aim at solving the issues at the state level, like sovereignty or territorial integrity, science research cooperation in the South China Sea aims at more down-to-earth purposes, namely ensuring that fishers can fish safely, that marine products for human beings are unpolluted, or that marine resource are protected in the right way.

Dr. Chu Manh Trinh, a Vietnamese senior expert in marine science, states that "the coral reefs in Paracel and Spratly islands need to be carefully protected for the whole East Sea Region, that is not only for the life of fish but the life of people in East Sea countries and the world. We need to have more dialogues and cooperation to protect and conserve these natural resources for human beings."[25] Therefore, the social consensus for this type of diplomacy will be higher than any others. Moreover, science diplomacy can attract additional intellectual capital from other sectors of society. As already mentioned, it is among a few types of diplomacy where non-state actors can participate. This means that every class in society, especially the non-officials—like scientists and staff of multilateral agencies—who would not normally be able to sit together at an official diplomatic table, can join in without making concessions, and without worrying about how what they say plays out with various political interests at home. This will facilitate a peaceful environment in which the parties are able to engage in "inventing without committing,"[26] which can yield creative ideas of new diplomatic paradigms for advancing intractable South China Sea issues.

For example, Vietnam welcomes NGOs to participate in environmental problem-solving initiatives. While the civil activities related to the South China Sea dispute management in political terms is quite limited and not openly public, environmental organizations like the Center for Development of Community Initiative and Environment,[27] Mekong Delta Youth,[28] Marine life Conservation And Community Development,[29] and others whose activities are aimed at solving the environmental issues in South China Sea, and have been working professionally with the participants of all classes in society, including young people, fisheries, businessmen, and others. In that sense, environmental advocacy translates into successful diplomatic efforts.

Second, now is the right time for science diplomacy to significantly exert its effect on the South China Sea dispute settlement. Since the Hague ruling affirmed that China has no legal basis with which to claim historic rights to the contested sea, China has not accepted the tribunal's decision. Thus, the power of the legal card in this conflict seems to be significantly decreased while the strategic and security aspects still remain unchanged or even more and more important. As a result, the sovereignty disputes have become tenser than ever with the dramatic rise of nationalism. After the tribunal ruling, with the exception of China, other claimants in the South China Sea, especially the Philippines, are faced with the challenge of enforcing the ruling. There is international and domestic pressure to show respect for international law, but they also need to find the best way to deal with China to avoid an armed conflict, which can easily take place if parties are unable to "exercise restraint and not act in a way that will raise tension".[30,31]

Although China has clearly shown its displeasure toward the ruling, it still seeks to be accepted on the world stage as a responsible international stakeholder. "The most promising outcome for all concerned would be a face-saving climb-down by China. Under this scenario, Beijing would promote détente rather than confrontation—without explicitly abandoning its jurisdictional [claims]," says Dr. Stewart Patrick, a senior fellow and director of the program on International Institutions and Global Governance at the Council on Foreign Relations. In that case, an approach that shows adherence to international law, especially with respect to UNCLOS, but also allows China to save face, would be certainly ideal. Science diplomacy seems to meet all of those conditions. This approach fully expresses the spirit of "marine scientific research" promotion regulated in Part XIII of UNCLOS.[32] It also provides a less sensitive way of interpreting the South China Sea disputes, and focuses on the resource and environment security issues where all the claimant parties can easily have a common voice and reach agreement. Therefore, it seems that now is the perfect time for science diplomacy to be widely exercised across the South China Sea region.

Third, science diplomacy has succeeded before in managing South China Sea disputes and in settling conflicts in general. In fact, science diplomacy is not a fresh idea in managing resolution conflicts. Its advantages have proved to be successful many times. For example, the Joint Verification Experiment (JVE) in 1988 between the US and the USSR during the Cold War was seen as "path breaking at its time, and it set the gold standard for government-to-government efforts."[33,34] The JVE was a US-Soviet collaboration to measure the explosive yields of nuclear tests by each side in order to provide a verification mechanism for a treaty to limit such testing yields. This is an initiative to solve the tension rising after the ratification of the 1974 Threshold Test Ban Treaty between the US and the USSR because of their mistrust about other side's nuclear weapon yield measurement methods.

In the context of the Cold War, this initiative had a significant meaning not only because of its contribution to keeping the US-USSR nuclear control agreement alive but also in the sense that it proved the ability of science diplomacy in initiating the cooperation between the two rival sides during a tense time. Viktor N. Mikhailov, a leader of the Soviet technical delegation to the JVE at the Nevada Test Site, claimed that "I am certain that the main result of the Joint Verification Experiment was not the development of procedures and extent of nuclear test monitoring of the joint development of technical verification means, but the chance for interpersonal communications with the American nuclear physicists."[35]

In addition, science diplomacy in the South China Sea also generates positive effects on the dispute settlement process when it attracts more and more interest from claimants with demonstrated successes. For instance, The Informal Workshop on Managing Potential Conflicts in the South China Sea has been continuously organized since 1990 with an increasing number of meetings and members.[36] In 1994, presidents of the Philippines and Vietnam signed a bilateral agreement to conduct Joint Oceanographic and Marine Scientific Research Expeditions in the South China Sea (JOMSRE-SCS). After 11 years of scientific research, the findings on marine biodiversity showed that the Spratly Islands could be a source of coral propagates for destroyed reef areas in the southern and western Philippines. However, the densities of marine species associated with offshore coral reefs were found to have been drastically reduced, particularly in shallow waters where blast and poison fishing are common. This project not only provided strong evidence that heavy exploitation of the fishery resources has occurred in the South China Sea, but also demonstrated a cooperative governance mechanism for larger-scale research and conservation. The JOMSRE-SCS is about to move into a second phase that goes beyond bilateral cooperation to a multilateral one, with the participation of China and all member states.

This is a notable achievement of science diplomacy in addressing the South China Sea disputes. It could bring all the claimant nations together to share the perception that scientific cooperation works effectively towards the common good for the entire region. This gain serves as an example for how science diplomacy can initiate activities contributing to the management of conflict. Countries in the region have also taken steps to establish their marine protected areas and gain certain successes. Since 2010, Vietnam has embarked on an ambitious initiative to create "national marine protected areas." As a result, the country currently has eight established with plans to add eight more in the near future. The goal is that this state-led environmentalism will create a transformative mindset among the nation's younger citizens and their relationship to the sea.[37] These incremental successes demonstrate compelling evidence that science diplomacy has the credibility to become an effective approach for South China Sea disputes settlement.

Finally, science diplomacy related initiatives can secure the support of major powers. Although this approach cannot fully satisfy the ambitions of any major power in the South China Sea disputes, it does bring them certain benefits without any negative effects. Science diplomacy is the most satisfactory explanation for the presence of powers like the US in the region.

However, the South China Sea dispute is, ultimately, a regional conflict where the claimant nations are the direct stakeholders. Therefore, outside interventions, either military or economic, are sensitive and can be misinterpreted in many circumstances and ignites nationalism, leading to negative reactions from some leaders or political groups among claimant nations. Filipino President Duterte's statement against US military presence in this country can be taken as an example.[38,39] In this case, science diplomacy can be used as a valid justification for the US and other non-claimant states to participate in the process of settling the dispute without being criticized for "interfering" in the other countries' affairs. Furthermore, science diplomacy provides China a way to get along with other claimant nations while maintaining its own dominant position on the issue. In other words, science diplomacy offers China a significant assistance in improving, or at least not deteriorating, its relationship with other claimants in the region without compromising its rhetoric about its "peaceful rise." Nevertheless, recent developments reveal protests against China's aggressive moves in the South China Sea from countries that seem to be China's friends like Malaysia,[40] a claimant, and Indonesia, another involved party.[41] Whether China wants it or not, its actions pose a dangerous threat to every state in the region, which most certainly presents obstacles to China's ambition to be recognized as a great power regionally and globally. Beijing's scientific cooperation with other states to solve the issues that are widely considered "practical and objective" in a peaceful manner, will help this nation partly erase their hegemonic image.[42] Or, as suggested by Professor Kathleen A. Walsh at the US Naval War College, it can be seen as a "carrot" offered by the Chinese government after a lot of "sticks" they have been posing in the South China.[43]

Lawrence E. Susskind, a Ford Professor of Urban and Environmental Planning at MIT, also raises the issue of China's support for science diplomacy. In an email communication, he found out that although China did not join some science diplomacy initiatives, they do have their own agencies for dealing with environmental issues. For example, China is not among the states that share data with the Pacific Tsunami Warning Center (PTWC) in Hawaii that collaborates with countries all over the globe in tsunami warnings, but has established its own tsunami warning center in the South China Sea— that is counted among its diplomatic initiatives in its SCS activities. China may already have their own science policy in the South China Sea that can prove both diplomatic and controlling at the same time.[44] As a result, they will be likely to at least not protest other similar initiatives. Also, with its huge economic advantage, China, to some certain extent, can dominate scientific initiatives to position its role as a firm supporter for environmental issues. For those reasons, it can be acknowledged that science diplomacy will get some support, or at least no hindrance, from major powers to forge an initiative toward the dispute settlement in the South China Sea. After all, scientific collaboration aims at a constructive resolution of common problems, and from all accounts, those problems related to environmental security are not going away any time soon. The real promise of science diplomacy is for nations and their citizens to recognize that the South China Sea is part of the global commons, and that it falls on all of us to conserve, protect, and sustain it.

The immense biodiversity that exists in the South China Sea cannot be ignored. Marine biologists, who share a common language that cuts across political, economic, and social differences, recognize that the structure of a coral reef is strewn with the detritus of perpetual conflict, and it represents one of nature's cruelest battlefields.

A joint scientific declaration for urgent action on an environmental moratorium on dredging is much needed. Recent biological surveys in the region and even off the coast of China reveal that the loss of living coral reefs presents a grim picture of decline, degradation, and destruction.

Professor John McManus, a recognized ecologist and coral reef specialist at the University of Miami, has regularly visited the SCS region and provided analysis to the tribunal. He has stated that based on satellite imagery analysis, Chinese dredging and clam poaching reveals a disturbing pattern of ecological destruction. McManus has researched this region for more than a quarter of century. He has called repeatedly for the development of an international peace park and is hopeful that other regional marine scientists and ecologists in this contested region will support this collaborative science-driven initiative.

Science Diplomacy—Limitations and the Way Forward

Given all the advantages mentioned above, the exercise of science diplomacy in the South China Sea dispute also meets with some limitations. Science diplomacy can be resisted by the rise of nationalism. The prerequisite for the success of science diplomacy is the cooperation among many nations without being influenced by territorial claims. However, in some cases, this effort can be interpreted as the compromise of the government in terms of sovereignty issues that can easily fuel nationalist outcry among claimants. For example, after Joint Marine Seismic Undertaking (JMSU) agreement among Vietnam, the

Philippines, and China to jointly explore the EEZ for the huge potential oil and gas reserves that are widely believed to lie under the South China Sea was signed in 2014,[45] there was a wave of opposing protests claiming that participation of the Philippines in this agreement translates as a "sellout of national sovereignty to China" because it undermined any future sovereignty claim by putting aside the territorial conflicts in the joint exploration.[46]

As a result, the JMSU lapsed in July 1, 2008, and it was not extended after receiving a lot of criticism from signed parties, especially the Philippines.[47] Also, science cooperation, which is open to the participation of NGOs, is sometimes viewed as a threat to nationalists. In fact, the questions about the relationships between international organizations, or NGOs in general, and state sovereignty have been under discussion for a long time. This is especially true among smaller, weaker states who are considered "the most frequent targets of external efforts to alter domestic institutions."[48]

As mentioned, South China Sea nationalism at times rises dramatically like a spring tide. Therefore, even a solution that seems to be completely neutral like science diplomacy can also be a source of hostile moves. It poses real challenges for easing nationalist sentiments associated with territorial disputes.

According to Karin Dokken, a political scientist at the University of Oslo, "The states around the South China Sea are to a large degree interdependent when it comes to questions of the human environment. They are interdependent to the degree that if they fail to find common solutions to environmental problems, they may end up in violent conflict and a potential for international integration."

In fact, China, as a major power and claimant in the South China Sea, its attitude is one of the most decisive factors to the success for every initiative to the dispute. The arguments about the support of China to science diplomacy options is built on the assumption that China would continue to follow its "peaceful rise" policy, which was re-emphasized by the current Chinese President Xi Jinping in 2014. He stated, "There is no gene for invasion in Chinese people's blood, and Chinese people will not follow the logic that might is right. [...] China will firmly stick to the path of peaceful development."[49]

However, with China's continued provocative activities, namely land reclamation in the disputed Paracel and Spratly Island chains,[50] this statement cannot be seen as a comprehensive assurance. Consequently, whether China would be supportive to science diplomacy initiatives, should be taken into consideration. Given that China has far greater capabilities than other states and can permit fishing or prevent fishing anywhere in the South China Sea, science diplomacy seems to serve the interests of smaller states more,[51] which is not a preferable scenario for China in pursuing its ambitions in the South China Sea. If science diplomacy can not find the way to prove itself as a source of tangible benefit to China's will in the South China Sea dispute or in the entire regional security architecture, its possibility to be realized is quite low.

From science diplomacy's challenges and opportunities outlined, this chapter offers some possible options that can be adopted by marine scientists and policy shapers to manage the South China Sea disputes:

- Establish complete freedom of scientific investigation in contested atolls and reclaimed islands;
- Expand scientific cooperation among all ASEAN marine scientists through more academic workshops;
- Provide an ASEAN regional cooperation science framework that mobilizes countries to address trans-boundary issues;
- Set aside all territorial claims;
- Invite ASEAN environmental NGOs to participate;
- Create a regional Marine Science Council to address environmental degradation issues;
- Foster dialogue for a proposed marine peace park;
- Propose a science-led ASEAN committee to study the Antarctica Treaty and the United Nations Environmental Program initiative under the East Asian Seas Action Plan;
- Organize devising seminars to advance collaborative problem solving in complicated policy disputes which bring together representatives of core stakeholders to brainstorm plans or strategies in unofficial conversations.[52,53]

Although science diplomacy is not a completely new approach to solving conflict in the South China Sea, the adoption of this peace-building mechanism by all claimant nations in the dispute is now urgently needed more than ever. This need is underscored by Lawrence Susskind and Saleem H. Ali in their pioneering book, *Environmental Diplomacy Negotiating More Effective Global Agreements*, in which they insist that now is the time for scientific advisors to demonstrate roles as "trend spotters, theory builders, theory testers, science communicators, and applied-policy advisors."[54] With ecological politics steering this South China Sea narrative, science diplomacy offers hope in protecting coral cathedrals, marine habitats, and fish species. It not only effectively addresses the disputes in the South China Sea but also is a possible peace building key for other similar environmental conflicts.

Appendices

Appendix A

Vietnamese Environmental Nationalism and the Campaign to "Save the East Sea,"

Jonathan Spangler and James Borton/Perspectives, 8/2016

The South China Sea Think Tank interviews James Borton about his experiences in the Chàm Islands, the emerging environmental awareness there, and the campaign to inspire Vietnamese youth to become more involved in national environmental issues.

James Borton is an educator and environmental writer. He is working on the book *Dispatches from the South China Sea* about the East Sea and Vietnam's ancestral fishing grounds, boat builders, and fishers. He recently traveled to the marine protected Cham Islands. He is the editor of *The South China Sea: Challenges and Promises.*

South China Sea Think Tank (SCSTT): You've been writing about Vietnam for more than a decade, focusing in recent years on the coastal and maritime environment. What sparked your interested in these issues?

James Borton (JB): I am long-term waterman, sailor and an ocean steward. Our oceans are under attack from climate change, El Niño weather patterns, sea surges, ocean acidification, and the world's voracious appetite for fish. Access to fish stocks is vital not only for Vietnam's fishermen but for many coastal nations. The East Sea [as the South China Sea is referred to in Vietnam] is the lifeblood of the nation's trade and is one of the world's richest fishing regions. It has always served as Vietnam's traditional, if not ancestral, fishing grounds. The future will define Vietnam's dilemma if the present young generation does not take steps towards conservation and sustainability of the dwindling marine resources.

SCSTT: You recently traveled to the Chàm Islands off the coast of Vietnam. Why there?

JB: I was invited by Dr. Chu Manh Trinh, who accompanied me to Cu Lao Cham. For the past eleven years, this youthful and engaging 53-year-old marine scientist has been traversing the twenty kilometers of open sea from Cua Dai Harbor in the city of Hoi An to ensure the success of this marine protected area and to preserve this paradise. When he first arrived, the islands were littered with nylon bags, trash, and sea animal carcasses. I learned that he embarked on a long-running mission to improve the environmental awareness of locals and enhance the islands' allure for tourists. Wherever we strolled on the island, the locals enthusiastically greeted him as the "professor," demonstrating respect but also giving the impression that he was part of the family. From crab to fish catches, Trinh has provided a "nautical chart" for their future by educating them about conservation and sustainable practices.

SCSTT: Can you tell us some more about this marine protected area?

JB: Vietnam has planned and approved many marine protected areas. While many have either failed or only partially achieved their management objectives, I witnessed firsthand this successful eco-tourism model revealing a rainbow of tropical life lurking among the hard and soft corals. On this pristine island, which you can get to by boat from the harbor in Hoi An in about 25 minutes, I met with local fishermen who understand and have been practicing sustainable fishing and habitat conservation. All of the island's residents know that the East Sea and their coastal waters is their safety net for life. Visiting the island, I was reminded of something I read in a book called *The Ocean of Life: The Fate of Man and the Sea* by marine biologist Callum Roberts, where he says, "It is essential for ocean life and our own that we transform ourselves from being a species that uses up its resources to one that cherishes and nurtures them."

SCSTT: What can you tell us about this island and what it represents for Vietnam's future?

JB: In speaking with Kinh fishermen, like sixty-year-old Nguyen Qui Hien, about their cultural beliefs associated with whales giving protection to fishers, I better understood the term "blue mind." In a water-based community, people are in daily harmony with the elements and often imbue their environment with magical qualities. Here on this island, there's not competition among fishers but cooperation. I saw them fixing their fishing nets, working on their boats, and in the market selling their fish. This ancestral fishing village is now a biosphere reserve, conserving and protecting the coral reefs, ecological diversity and cultural heritage. This sense of their ancestral past is reflected in the many pagodas and temples dedicated to honoring whales. In their beliefs and festivals, the islanders reveal their pride in their livelihoods and, more importantly, their community. The Chàm Islands are indeed a breath-taking cluster of granite islands, and the livelihoods of the population depend on fishing and other marine resources.

SCSTT: It sounds like the people of the community may have some lessons worth sharing. What kind of things do you feel could be learned from them?

JB: I think that they can teach all of us about the need to conserve and to sustain our oceans. Dr. Trinh and others at his marine protected area have taught the Kinh that by protection of their coral reefs, they are saving their seas for more fishing for the future generations. The islands boast 277 coral species and 270 reef fish species as well as a wealth of natural beauty and an abundance of traditional knowledge and customs. Even though it is becoming a tourist destination, there is no development and no hotels. The community has brought in a new form of income in the way of homestays for divers and eco-tourism. I have also learned from them how people can truly have a deeper emotional connection to the sea. The East Sea inspires, thrills, and even soothes people. I encourage Vietnam's younger people to visit this magical island so that they can experience how this body of water, the East Sea, casts its spell and holds us in a net of wonder forever. I know that I will return soon.

SCSTT: Having worked as an educator, how would you suggest getting youth interested in maritime environmental issues in the area?

JB: Why not start a social media-inspired campaign among Vietnam's net-savvy citizens? A few people have suggested calling it "Save the East Sea" or Lưu giu Bien Dong. It is becoming apparent that Vietnam's maritime history can be seen in every cultural and economic activity found in the nation's ancestral fishing grounds, traditional boat building industry, and marine protected areas. What Vietnamese are witnessing is the continuity of tradition in opposition with rapid cultural changes. Perhaps young people can create a national conversation about some of these issues related to East Sea ancestral traditions and the challenges of sustaining them.

SCSTT: Why might this subject or message on social media interest young people?

JB: Well, for starters, the contested sovereignty claims are not going away any time soon. Furthermore, Vietnam's fishing communities are central to any examination of sovereignty and should be placed into the national discourse as part of a nationwide "soft diplomacy" response to China's continued aggressive actions in the East Sea. Finally, overfishing and pollution have immediate and direct impacts on their lives now and tomorrow.

SCSTT: Interesting. So, in essence, are you saying that increased interest in the South China Sea disputes might actually inspire people to pay more attention to environmental issues?

JB: Yes, the world has seen the startling satellite images. This body of water is one of the most important large marine ecosystems in the world, rich as it is in marine resources from fish species and coral reefs. Hundreds of millions of lives depend on it as a food source, so it's no wonder that more marine scientists are weighing in on the environmental catastrophe associated with China's reclamation since it endangers fish stocks and kills coral reefs.

SCSTT: If you could tell people there in Vietnam and elsewhere one last thing about these issues, what would that be?

JB: Take charge and control of your future. Start a campaign to conserve and to sustain the East Sea. Plan a ceremony the first week in June 2016 and hold it on Cua Lao Cham. The program might even correspond with a scheduled 2 to 3-day real time and online writing workshop on "Understanding Vietnam's Environmental Challenges in the East Sea." All this could become a grand lead up to World Ocean Day to be held on June 8, 2016. During this week, throughout Vietnam's coast, young people can engage in a sweeping cleanup of the coastal environment and disseminate government-approved information on the environment and how to "save the East Sea." All these efforts may succeed in mobilizing and uniting Vietnam's population to claim and to protect their precious and fragile ancestral fishing grounds. In the best of worlds, a Vietnam social media campaign would encourage a form of environmental nationalism, and maybe it could even inspire other neighboring countries to promote conservation and sustainable development of the maritime environment as well.

Jonathan Spangler is the Director of the South China Sea Think Tank and a Doctoral Fellow at the Institute of European and American Studies at Academia Sinica in Taipei, Taiwan.

Citation

Spangler, Jonathan, and James Borton, "Interview with James Borton: Vietnamese Environmental Nationalism and the Campaign to 'Save the East Sea.'" *Perspectives*, vol. 8, 2016. Taipei: South China Sea Think Tank. http://scstt.org/perspectives/2016/538/.

The South China Sea Think Tank serves as a platform for promoting dialogue and does not take any institutional position regarding maritime territorial claims. Published material does not necessarily represent the views of the organization or any of its individual members. While the SCSTT makes every attempt to provide accurate information, contributors are solely responsible for the content of their own articles.

"Science Diplomacy" as a Solution to the South China Sea Disputes?
By Jonathan Spangler and James Borton/Perspectives, 1/2015

The South China Sea Think Tank interviews James Borton about "science diplomacy," prospects for international cooperation on environmental issues, and Taiwan's role in the South China Sea.

James Borton is a veteran environmental policy researcher, former foreign correspondent for The Washington Times, editor of the book The South China Sea: Challenges and Promises, and a non-resident research fellow at the Saigon Center for International Studies (SCIS) at the University of Social Sciences and Humanities, Ho Chi Minh City, Vietnam.

South China Sea Think Tank (SCSTT): "Conflict." For many, that's the first word that comes to mind when they hear "South China Sea." What do you think of?

James Borton (JB): Rival nations have wrangled over this territory for centuries, but China's economic and military rise have sparked concern about their control of atolls, islands, sandbanks, and reefs. Never mind that this sea is a major shipping route and home to rich fishing grounds that offer a cheap source of protein for people. Robert Kaplan rightfully calls the South China Sea "the 21st century's defining battleground." It's no wonder that other claimants, especially Vietnam and the Philippines, have engaged in recent clashes with China's navy and fishing vessels over disputed territorial claims. CNN's circulated images of China's rapid military modernization and assertive behavior in the SCS cannot help but translate into renewed fears of conflict.

SCSTT: What would you say motivates claimants?

JB: China's completed reclamations on eight maritime features in the South China Sea cannot be explained by either basic economic motives or by the important but declining fishing stocks around the Spratly Islands. SCS surveys reveal that the oil and gas reserves are largely insignificant and coupled with the high cost of deep-water drilling and the propensity of political risk render energy extraction unprofitable. In fact, most of the natural resources lie outside the areas affected by China's artificial island-building. The reality is not that there is unexplored gas and oil fields under the Spratlys but rather it's the tankers filled with oil and gas that floats past them. More than half of the world maritime trade passes through the SCS with the globe's busiest shipping lane passing right by the Spratlys. Control this flow and China controls the energy security of Asia. This leads to only one conclusion: China's has buttressed its claims in the disputed waters and features 700 [nautical] miles from its coast to fortify its military foothold in the South China Sea. The construction of a runway on Fiery Cross Reef and the Spratlys is part of a

Chinese military strategy that includes bolstering its naval strength beyond the mainland and into the open seas.

SCSTT: In our previous conversations, you've talked about the idea of "science diplomacy." What first sparked your interest in that?

JB: After meeting Professor John McManus, a South China Sea coral reef specialist from the Rosentiel School of the University of Miami, and learning about his pioneering efforts to generate support for a marine protected area in the Spratlys as early as 1992, I thought that his important science-driven environmental policy needed to be examined more fully to address the mounting SCS political and environmental problems. The problems are disturbing. Nearly 80 percent of the sea's coral reefs have been degraded or under serious threat in places from sediment, overfishing, destructive fishing practices, pollution, and climate change. Recent biological surveys in the region and even off Mainland China reveal that the losses of living coral reefs, present a grim picture of decline, degradation, and destruction. The scientific community supports the overwhelming evidence that China's continued reclamation of atolls and rocks through the dredging of sand in the Spratlys disrupts the fragile marine ecosystem. The area has been recognized as a treasure trove of biological resources and is host to parts of Southeast Asia's most productive coral reef ecosystems. It seems that there is a developing consensus among SCS marine biologists that their science needs to be communicated to policymakers now, or barring that, the larger issues of food security and indiscriminate destruction of coral reefs will lead to an ecological catastrophe. More scientists like McManus appear to be joining the ranks to alert the policy shapers and general population that science diplomacy is the glue for building constructive collaboration and partnerships.

SCSTT: Your recent book, *The South China Sea: Challenges and Promises*, has contributions from a great team of top South China Sea scholars, including quite a few representing the Vietnamese perspective. What do you hope will be the main takeaway for readers of the book?

JB: That China's expansionist behavior offers an array of challenges not only for Vietnam and ASEAN but also for Washington. The international community cannot accept that one nation has the right to nationalize the open sea for its strategic purposes. The sea and all of its marine life is transnational and so, by its very nature, transcends national maritime jurisdictional boundaries. The sea will be there long after all sovereignty issues disappear over the horizon.

SCSTT: What's the current state of cooperation on science and environmental issues in the South China Sea?

JB: Professor Hai Dang Vu from the Diplomatic Academy of Vietnam and Professor Aldo Chircop, former Chair in Marine Environment Protection at the IMO's World Maritime University in Malmo, Sweden, both environmental law experts, have proposed marine protected areas (MPAs) in both the contested and uncontested areas of the SCS. They believe that these systems can be tailored to address the specific structure, function and processes of its large marine ecosystem as they may be defined in spatial terms. I believe they are calling for a special designation as a particularly sensitive sea area under the IMO framework for added protection. All this points to the generation of a possible network of MPAs. Vietnam conducted regular Joint Oceanographic and Marine Scientific Research Expeditions in the East Sea in cooperation with the Philippines from 1996 to 2007. These experiences have contributed to forming a cooperation model on marine science research. After all, all good science is a collaborative exercise. What it offers is validation for a shift in interactions between scientists, resource managers, and policy makers through international marine science partnerships. As a footnote in marine research cooperative models, it's noteworthy that China and Vietnam collaborated in a Comprehensive Oceanographic Survey in the Gulf of Tonkin in 1959–60 and 1962, where large numbers of specimens of marine fishes and invertebrates were collected and deposited in the Marine Biological Museum of the Chinese Academy of Sciences in Qingdao for taxonomic and biodiversity studies. South China Sea Monsoon Experiment (SCSMEX) was initiated in 1998, which was an international field experiment with the objective to better understand the key physical processes for the onset and evolution of summer monsoons over Southeast Asia and southern China aimed at improving monsoon predictions. The scientists were from Taiwan, Australia and America.

SCSTT: How does environmental policy fit in amidst all the political and diplomatic tensions in the region?

JB: It is the centerpiece in this complicated South China Sea chessboard. After all, the United Nations Convention on the Law of the Sea (UNCLOS) entered into force in 1994 and presently has 152 parties, although US ratification is still pending. There are at least two vital parts of that law and maybe forty specific articles that directly apply to marine scientific research or at least the development and transfer of marine technology. Here are a few of the relevant articles: the promotion of marine science and technology capacity building, particularly in developing countries; the encouragement and facilitation of international cooperation in marine scientific research and development; and the establishment of regional marine science and technology centers. In short, the lessons learned from other marine-based scientific collaboration point to improved communication and working partnerships between seeming adversaries. This focus on the environmental challenges enables a new conversation or, better yet, the shaping of a compelling narrative with a strong educational outreach. This year Vietnam and the Philippines have reaffirmed their maritime cooperation focusing as much on shared fishery data, marine science research and marine environment protection. Science ability, the role to integrate research and the collaborative monitoring of data can only boost geopolitically informed decisions. These interactions among scientists, policy managers, and the public, now shaped effectively and instantaneously through social media, help direct and define strategies for peaceful co-existence in the fragile management of precious marine resources.

SCSTT: Shallow waters often serve as sanctuaries for biodiversity. What kind of marine life are we looking at around the islands and sea features of the South China Sea?

JB: There have been approximately 1,787 fish species recorded within the South China Sea; however, only a few of these are endemic to this sea. Only a handful of mostly tiny islands, atolls, and reefs—the Spratly Islands to the south, the Paracel Islands to the north—break up the largely featureless maritime plan that separates Vietnam from the Philippines along the east-west axis and Hong Kong from Borneo from north to south. Taiping is the largest of the Spratlys at 1.4 kilometers in length and 0.4 kilometers wide. Despite its inconsequential size, it's the only one with its own freshwater. Marine biologists and taxonomists may want to undertake a systematic survey to determine how many multi-cellular species are located specifically near the coral reefs. Vietnamese fishermen tell me that the major commercially available fish species found are yellow croaker, filefish, chub mackerel, Chinese herring, and various species of shrimps. Although the number of fish species in the South China continues to decline, the remaining coral reefs contain more unique sea creatures. The variety of species living on a coral reef is greater than anywhere else in the world. It's estimated of 70–90% of fish caught are dependent on coral reefs in Southeast Asia and reefs support over 25% of all known marine species. The problem is that these coral reefs are being destroyed daily.

SCSTT: What environmental damage and threats are we facing here?

JB: Unfortunately, fragile coral reefs are now threatened by ocean acidification, overpopulation, overfishing, reclamation, sedimentation, and destructive fishing practices. Even China recognizes that their coastal waters have been wrecked by rapid industrialization and also the need to transplant coral reefs wrecked by reclamation damage. All this happened despite China's passage of the Marine Environment Protection Law in 1982. Most marine scientists agree that China's marine environments, especially in the South China Sea and Yellow Sea, are among the most degraded marine areas on earth. Loss of natural coastal habitats due to land reclamation has resulted in the destruction of more than 65 percent of tidal wetlands around China's Yellow Sea coastline. The dredging on these SCS atolls kills reef formations with the dumping of sand and concrete. Furthermore, the placement of personnel on these reclamations only succeeds in bringing sewage, garbage, marine debris, oil and gas spillage into the once pristine waters.

Marine scientists have made it clear that many populations of reef fish do not migrate and mix with others across oceans. Instead, new studies suggest that larvae tend to settle near where they were born. So many species of fish exist in small geographic ranges and destroying even one small section of reef does lead to extinction.

SCSTT: We've learned from your articles that some of the fishing going on here isn't your basic nets and fishing lines. Could you tell us a bit more about what kind of fishing operations are taking place?

JB: For sure, the rapid rise in marine catches has been largely brought about by the significant increase in fishing vessels and, of course, the size of these trawlers and their mega-holding tanks. China has over 30,000 commercial fishing trawlers engaged mainly in bottom trawling and some in the use of gill nets. Vietnam has at least 20,000 commercial fishing trawlers that are soon to be replaced by steel hulled vessels with larger capacity since the government has now made available subsidies in the form of low-interest loans to modernize their colorful wooden boats. It's easy to see that, with the increased number of fleets from the Philippines, Indonesia, Malaysia, Japan, and Taiwan, there's a modern fishing war unfolding. These super trawlers with their modern technology, such as remote sensing, sonar and Global Positioning Systems, together with incentives and subsidies, have brought high production from deep-water and high sea areas and habitats, such as continental slopes, seamounts, cold-water coral reefs, and deep-sea sponge fields, into the reach of fishing fleets trying to exploit the last refuges for commercial fish species. Fishing vessels are now operating at depths greater than 400 meters and sometimes 1,500 to 2,000 meters deeper. With fish catches plummeting, fierce competition among fisheries in the region contributes to more open conflicts between fishing fleets. Perhaps the most worrisome fisheries problems in the South China Sea are the destructive practices of dynamite and cyanide fishing. This widespread practice of the use of dynamite is found from Indonesia to southern China and even off Vietnam's coasts. This deleterious practice typically occurs on or near coral reefs causing further destruction of the fragile ecosystems. Gill nets are used at all levels of fishing, from commercial scale to the family or personal scale, throughout the SCS. Gill nets are designed to allow the fish to begin to swim through but trap the fish at the gills, preventing their escape. Also, the use of these nets is a major threat to sea grass beds.

I am reminded of what Paul Greenberg wrote: "To be sure, the postwar assault on fish sprang from an honorable intention to feed a growing population that boomed in a prosperous postwar world. But as in war, everybody loses when there is nothing left to fight for." It's clear to fishermen since they are the world's sentinels for our fish supply that marine fisheries represent a significant, but finite, natural resource for all coastal nations.

SCSTT: Historically, many attempts to promote international cooperation on environmental issues have struggled to achieve their goals. Is there any precedent for international environmental cooperation in the South China Sea?

JB: Antarctica is the one place that arguably is the archetype for what can be accomplished by science diplomacy. Under the Antarctic Treaty, no country actually owns all or part of Antarctica, and no country can exploit the resources of the continent while the Treaty is in effect. It is a classic example of international cooperation. Over time, the Antarctic Treaty developed into the Antarctic Treaty System, which includes protection of seals and marine organisms and offers guidelines for the gathering minerals and other resources. Additionally, the Arctic Council has been able to effectively steer the passage of domestic legislation, international regulations, and, most importantly, international cooperation among the Arctic States. Eight nations—Canada, Denmark (Greenland), Finland, Iceland, Norway, Russia, Sweden, and the United States—have territories [claims] in the Artic, and the domestic laws of these nations govern actions taken within their territorial waters. Also, it's worth noting the success of the Red Sea Marine Peace Cooperative Research, Monitoring and Resource Management Program (RSMPP) where Israel and Jordan signed off on an ecosystem monitoring agreement and shared science data collection in the Gulf of Aqaba in 2003. RSMPP offers a model for improving international relations and building capacity through marine science cooperation in the South China Sea. These two opposing countries chose to promote the long-term sustainable use and conservation of their shared marine resources.

SCSTT: Even so, it's pretty clear that issues of territorial sovereignty are the current priority for littoral states. Why should governments care about environmental conservation?

JB: The escalating territorial dispute in the South China Sea is as much an ecological crisis as it is a geopolitical one. Dredging, land reclamation, and the construction of artificial islands appear to be swamping centuries old reefs in sediment, endangering ecosystems that play a key role in maintaining fish stocks throughout the region. If a nation can no longer feed its people, then there are riots, instability and war. Who has forgotten about the 2008 global food crisis that triggered food riots in more than 30 countries? The earlier call for a "Green Revolution" may have now been supplanted by a "Blue Economy" mantra, especially in the South China Sea. Climate change continues to show us that we live in a world without borders. This is even more certain about our oceans. Our leaders and scientists are examining how we think and act on a global scale to address the pressing environmental issues. Our oceans' health is fundamental to life on this planet. Marine scientists along with responsive and responsible governments realize that protection of a marine ecosystem may be the smartest investment of capital that we as a society can ever make.

SCSTT: What will it take to motivate regional actors to address these issues?

JB: The urgency is that the South China Sea is being overfished and pol-luted, and that's threatening the food supply of millions of people. I believe that the marine science message about the dire consequences associated with the wanton destruction of coral reefs, and the depletion of fishing grounds must be heard now by all claimant nation leaders. The fragile South China Sea marine environments must be managed collectively by all claimants or there will be nothing left to dispute. Perhaps marine scientists will summon Ministries of Education among all claimants to roll out an environmental message about "Saving the South China Sea." How about essay competitions engaging students from middle school to university to address this compelling and vital environmental issue? Social media networks coupled with NGOs can bring pressure upon all governments to respond to the environmental crisis. The answer is to educate the region's citizens and mobilize changing attitudes about the awareness the need for conservation of the ocean and the coastlines.

SCSTT: Would you say that governments are using the "environmental argu-ment" out of a genuine interest in protecting marine habitats in the area? Or is it more out of political and diplomatic self-interest?

JB: "Human beings are at the center of concerns for sustainable development." So, the Rio Declaration begins. All responsible and especially authoritarian governments know that the needs of their poor must be met. If the overfishing persists and the coral reefs are destroyed, then these SCS nations, rank and file, will experience the dire consequences of having squandered their resources, inequitably distributed wealth, and degraded their landscapes. People will take to the streets. "The centre cannot hold," claims W.B. Yeats, and he's so right. After all, governments, including China, no longer legislate population growth and so it is abundantly clear that, with dramatically escalating populations, a marine conservation policy surely must be adopted so that there will be resources for tomorrow.

SCSTT: You've suggested that Taiwan may have a key role to play in this. Tell us a bit more about that.

JB: Let's do examine Taiwan. There's wide recognition that the international marine research station on the Dongsha [Pratas] Island has succeeded in inter-nationalizing science research cooperation. In fact, both APEC and ASEAN acknowledge the impact of scientific collaboration has been achieved and is still ongoing. In conversations with Professor McManus, he emphasized that Dongsha's unique oceanography makes it an exceptional coral reef ecosystem that seemingly proves resilient to climate change. Also, it's most credible that a national park was established around Dongsha atoll. The research center, with its new lab facilities and even a research vessel, continues to attract marine researchers from the region to share studies and data. This center has literally become a crucible bringing together marine scientists who recognize Taiwan's focus on peace and prosperity, reconciliation, and cooperation. Although President Ma Ying-jeou's Peace Initiative was ignored by both China and Japan, it reframes and reinstates a much-needed idealism. I am tempted to equate his actions with a strand of Wilsonianism since his plan acknowledges "nationalism" while calling for a politically plural world. Taiwan is a responsible stakeholder in the South China Sea. It also confirms that any movement towards proper management of the SCS disputes should involve Taiwan since it controls the largest land feature in the SCS.

SCSTT: Why Taiwan? Why not another claimant, non-claimant, or international organization?

JB: Taiwan's complex cross-strait relationship with Mainland China underscores that sovereignty and security form the core of their relationship. Despite missed opportunities, missteps, and suspicions of bad faith, there is peaceful coexistence marked by negotiation and rapprochement. Under Ma's leadership, there have been at least 20 formal agreements inked with Mainland China offering prosperity in trade and social order dividends. Certainly, one can point to the cross-strait Economic Cooperation Framework Agreement (ECFA) as a success. Other SCS claimant nations can learn from Taiwan's example. Despite ongoing conflict management and confidence-building efforts in the South China Sea, there is still no clear path to the resolution of the complex multilateral sovereignty and the maritime boundary disputes. Intergovernmental Panel on Climate Change assessments for the region forecast significant climate and ecological change to the detriment of the region's coastal inhabitants, ecosystems, and economies. South China Sea nations need to place marine conservation cooperation at the center of all development activity in order to enhance the prospects of adaptation to climate change. With diminishing marine resources, all claimant nations are increasingly mindful of the need for sustainability. It's my belief that Taiwan's relationship with Mainland and the internationalization of Dongsha atoll reaffirm the nation's commitment to peace, conservation, and sustainability in the SCS.

SCSTT: How does your own research aim to contribute to all of this?

JB: While the intractable geopolitical SCS impediments remain, the Spratlys might be seen as a "resource savings bank," where fish, as trans-boundary residents, spawn in the coral reefs and encircle almost all of the South China Sea waters, before returning home. I hope to be invited to Taiwan as a visiting environmental writer to research why Dongsha is a crucible for scientific cooperation and to better understand a successful marine protected area. The Taiwanese government recognized Dongsha atoll's prominence as a model for the sustainability of fishery resources in the SCS and was designated as the first marine protected area in March 2004. There's much to be learned from speaking with marine scientists and policy shapers. The tipping point is that the assemblage of the South China Sea is increasingly shaped in scientific terms. By examining Dongsha, perhaps I can communicate why and how this scientific paradigm may be applied in other possible marine protected areas.

SCSTT: You've visited and spoken to people whose lives have been affected personally by the South China Sea disputes. What's one thing that really stood out for you in these conversations?

JB: It is about their sense of hope and resilience. The fishermen want to catch more fish and do not want to be attacked or rammed and sunk by other neighboring boats. The coastal populations want unpolluted waters and access to fish. Families want a better future. As I have suggested in my articles, there's a new conversation and narrative among marine scientists and policy shapers. All this points to science helping to make the case for joint development of marine resources.

SCSTT: To wrap up here today, as you know, the vast majority of information about the South China Sea focuses on the obvious problems and rarely on potential solutions. What opportunities would you say the future holds?

JB: Because of the scale of the destruction of coral reefs, the decline of fish stocks, the reclamation projects, there are increasing signs that claimant nations want to sign fisheries agreements, just as Japan did with Taiwan in 2013.

This again is an excellent model on how to preserve sovereignty claims in contested waters while taking the higher road toward sustainability goals. I fully expect ASEAN in 2016 to ratchet up its direction toward realistic environmental protection goals for the South China Sea. We will see more marine science driven joint working groups in 2016. It bodes well that next month in Haiphong, there's a major East Sea environmental conference. Stay tuned as the dialogues reach more policy shapers.

Jonathan Spangler is the Director of the South China Sea Think Tank and an Adjunct Lecturer with the College of Social Sciences at National Chengchi University in Taipei, Taiwan. His current research focuses on the effects of actor involvement on escalation and de-escalation in the South China Sea maritime territorial disputes. He has spent the past seven years living in East Asia.

Citation

Spangler, Jonathan, and James Borton. "Interview with James Borton: 'Science Diplomacy' as a Solution to the South China Sea Disputes?" *Perspectives*, vol. 1, 2015. Taipei: South China Sea Think Tank. http://scstt.org/perspectives/2015/442/.

The South China Sea Think Tank serves as a platform for promoting dialogue and does not take any institutional position regarding maritime territorial claims. Published material does not necessarily represent the views of the organization or any of its individual members. While the SCSTT makes every attempt to provide accurate information, contributors are solely responsible for the content of their own articles.

Appendix B

Science Diplomacy: A Crucible for Turning the Tide in the South China Sea
Audio transcription from East West Center Webinar
Held on Sep 30, 2020, 10:00am–11:30am

Dr Lina Gong is Research Fellow at the Centre for Non-Traditional Security Studies (NTS Centre), S. Rajaratnam School of International Studies (RSIS), Nanyang Technological University (NTU).

In East Asia, the concept of non-traditional security (NTS) has been frequently used since the beginning of the 21st century because of the 1997 Asian Financial Crisis and the SARS epidemic in 2003. NTS essentially refers to non-military problems that challenge the survival of the state and well-being of the people. Some of these problems include climate change, natural hazards, environmental pollution, food shortage, and infectious disease. The Covid-19 pandemic has shown to all of us what an NTS crisis can cause for states and individuals. The cost of such an event is no less or even more than a military conflict. Marine environmental pollution is obviously one of the NTS concerns in the region.

Under the rubric of environmental security, this issue receives increasing attention in ASEAN. This is evident in the adoption of the Bangkok declaration on combatting marine debris in the ASEAN region in 2019. National governments in the region have taken efforts too. The Philippine and Thai governments, for example, closed the popular tourist islands for environmental concerns. Because many of these issues are transnational, it is beyond the ability of any single or individual state to deal with it effectively. Management of these issues needs cooperation or joined efforts by the states concerned. And because of their non-military nature, they are considered less sensitive and therefore have been used as a channel to maintain dialogues and generate goodwill among countries. Japan and the US, for example, have been supporting ASEAN and its member states dealing with various NTS challenges such as natural hazards. China and ASEAN signed a joint declaration on NTS cooperation in 2002 and a few Memorandums of Understanding since 2004. Cooperation is an ideal and desirable way to deal with NTS challenges.

In the South China Sea, there are certainly cooperative activities, but it seems more should be done. In my view, the excessive emphasis on the state as the only security reference object is one of the many reasons that we do not have enough cooperation. So, when we talk about security, it is very important to specify whom or what we are protecting. Traditionally it is the state, which has sovereignty, territorial integrity, and population. It is from this angle that we are seeing competing claims in the South China Sea which have heightened tensions among the states concerned. Because territorial integrity is so important, the national governments are likely to alter their position. As a result, it is very difficult to substantiate or deepen cooperation in this body of water. From a non-traditional security perspective, the security referent object also includes the people. As the population increases, living standards rise, states are under pressure to sustain growth, to meet the growing demands of the population for food, water, energy, and other resources. Because of the limited nature of many of these resources—we have seen the depletion of resources and competition among states. Because of unsustainable economic activities such as IUU, excessive tourism, discharge of untreated waste into the seas, these environmental issues are getting more serious.

In the context of the South China Sea, we have seen reports of fishing disputes among the littoral states and such incidents are a source of tensions too. When traditional concerns entangle with non-traditional ones, the situation becomes even more complicated and concerning. In my view, environmental security defers from other types of NTS in a way that the environment itself can be a security referent object too. But this is largely missing from our discussions, particularly those focused on the South China Sea. We need to consider whether it is legitimate or justifiable to defend sovereignty, to meet human needs at the expense of the marine environment—including the sea creatures. What kind of relationship between the different security referent objects is healthy or sustainable?

Another practical way is to initiate or strengthen cooperation in issues indirectly related to the environmental status in the South China Sea–like education. Joint education programs between littoral states of the South China Sea or under ASEAN or marine science subjects, can be one option or a possible way for further cooperation. Or, cooperation in waste management in ASEAN member states. Much of marine plastic debris is actually from land-based sources and a lack of waste management facilities is an important cause for that. Given the growing interest among ASEAN member states to curb plastic pollution in the seas, joint research in these areas is also a possible pathway to the future. There are many barriers to cooperation in this body of water. To cope with or circumvent those barriers, the states may consider first, how they view the relationship between the state, people, and the environment. And second, start cooperation in the least controversial areas to build confidence.

Dr. Liana Talaue McManus is an independent marine scientist who is engaged in the Global Environment Facility Marine Plastics Project.

It is an honor to participate in this webinar. I'd like to share a few thoughts on how science may play a role in sustaining the South China Sea.

According to Dr. Jane Lubchenco, a former head of the US National Oceanic and Atmospheric Administration, we are supposed to have a new social contract for science because the world is changing very profoundly. Scientists need to lead the dialogue and scientific priorities, new institutional arrangements and improved mechanisms to disseminate and utilize knowledge more quickly. In this changing world, 193 nations have voiced their aspirations that they wish to pursue for their wellbeing and that of the planet. These are the 17 sustainable development goals that make up the sustainable development agenda to 2030.

In particular, I'd like to point you to Sustainable Development Goal #14, which is about preserving life in the oceans. It includes preventing and reducing marine pollution, conserving ecosystems such as coral reefs, mangroves, and sea grasses, managing fish harvests, and conserving at least 10 percent of coastal and marine areas.

A global assessment of transboundary large marine ecosystems, or LMEs for short, worldwide show that Southeast Asia have 5 LMEs. These LMEs, on average, are at high risk (the orange color), which is level 4 on a scale of 1–5 from low to highest risk. But what does high risk mean?

How does the South China Sea stack up against the aspirational goals? On the issue of pollution, nutrients and plastics are increasing and disrupting marine food chains. On marine protected areas, the South China Sea is 2/3 away from the minimum goal of 10 percent by area. On overfishing, around 40 percent of fish stocks are collapsed or overexploited.

Addressing Goal 14 for the ocean is linked to other goals. But how do we eradicate poverty among 38 million poor around the South China Sea, mostly fishing and farming families and urban dwellers? How do we eradicate hunger knowing that 38 percent on average of animal protein is supplied by dwindling fish stocks? Given environmental, economic, social, and political state of the South China Sea, what must scientists do?

When I went back to the Philippines in 1986 with John, there were more projects than graduate faculty. In the 80s, ASEAN cooperative programs in marine science were on hyperdrive. I participated in three that I show here. ASEAN-Canada focused on marine pollution and red tides. ASEAN-Australia focused on systematic assessments of coastal ecosystems. For ASEAN-US, it introduced the approach called Integrated Coastal Zone Management. These projects I call foundational and their legacies shape marine science in the region to this day.

The years following were a period of institution and coalition-building. The projects included implementing action programs and plans at regional, national, and subnational scales. Confidence-building measures focused on multilateral or bilateral platforms were taken, notwithstanding conflicts in jurisdictional claims. Given this rich scientific experience, what must scientists do to sustain the South China Sea? The revised social contact, under the sustainable development agenda, compels scientists to drive transformations on how we relate with one another so we can collectively protect, and can use the bounty of the South China Sea's large marine ecosystem.

There are four things that we need to do.

1. We must show that linked ecosystems, from land to sea, constitute our collective principle, in a changing and constraining climate. We must show how this principle must be managed to last generations, building scenarios, and communicating options for viable actions.

2. We must show that ecosystem health equates to human wellbeing. Degrading ecosystems is eroding the principle and it incurs very high costs for human wellbeing. This cost must be quantified and integrated in governments.

3. Human consumption of air, water, land, and sea must be underpinned by the rates at which ecosystems replenish. What happens under scenarios of optimal or excessive consumption, we need to quantify and communicate.

4. Finally, the asymmetry among nations can no longer be an excuse for unilateral actions. Only partnerships based on mutual and reciprocal trust and respect can safeguard the principle. We must all work hard and fast to nurture resilience in these partnerships despite tensions and disagreements.

Thank you.

Dr. Satu Limaye is Vice President and Director of the East West Center in Washington, DC.

Thank you very much indeed Dr. McManus, for a very eloquent and well thought out presentation on why we need to cooperate and particularly I was taken–I wasn't aware of the previous generation of joint projects. So immediately it came to my thoughts how we might be able to rebuild some of those projects. Of course, different times, different conditions, different environments, different needs. But in any case, the general principle of projects that build a regional human development capacity as well as a scientific baseline in a cooperative spirit which is the theme of this science diplomacy initiative.

I want to thank all of you for being remarkably well-disciplined in time. I'm not sure whether that is explained by your scientific backgrounds or a combination of your training, but I thank you because it's always tricky to do a program for 90 minutes on a webinar with so many distinguished speakers with so much to say.

Thank you, very few technical glitches and it's now my pleasure to turn over the proceedings to my colleague and collaborator who brought this program, as I mentioned at the outset, to our attention, and that is Mr. James Borton.

James, will you please take over the Q&A? And I see that some questions have already been placed in the queue. Over to you James and I will turn off my mic.

James Borton is an independent journalist and researcher who has been reporting on Southeast Asia for several decades and environmental security issues in the South China Sea.

Thank you very much. I want to thank the East West Center again and, of course, our distinguished panelists for their major contribution to today's webinar. I think all of us would have to agree with what our keynote speaker, Dr. Paul Berkman, stated in starting this program. He said, "Science diplomacy is the language of hope for humanity." And so, we've moved to our panelists, our scientists, who have really presented the clarion call for science to lead policy in addressing these dire environmental issues in the South China Sea. All of you have addressed climate change issues, ocean acidification, pollution from plastics, overfishing in the South China Sea, and of course, sadly, coral reef destruction. I think in opening this up, we are really focusing on three keywords: collective conservation, and protection of our ecosystems.

All of this acknowledges that cooperation in science and among nations is required to address the food security issue looming in the South China Sea. So, in terms of opening the questions up to the panelists, I wanted to frame some meta questions specifically.

What I'd like to do is frame a few general questions for all panelists rather than singling out individual panelists. The first one is that we're calling for this need for cooperation among the claimant nations to address how the role of science is necessary. So how do we convince claimant nations in the South China Sea to establish innovative ways to share their science data, to map out a blueprint for environmental conservation and a sustainability plan? The idea is that we want to have these scientists from all of the nations share their data and how relevant that is for creating a collective plan that will benefit all of those countries. What major steps can we take to create that kind of regional unity in the sharing of science data? Any of our panelists can address that and I welcome their responses.

1:08:11

Dr. Liana Talaue McManus: There is already experience in scientific data sharing—back with the ASEAN-Australia project, they established a reef monitoring system region-wide and the data is shared through a database. Albeit there may be levels of protection of the data, but by far they have adopted a policy of open data.

Mr. James Borton: Very good. Is there anyone else that would like to respond?

1:08:44

Dr. Paul Berkman: I'll jump into this, your question of innovative ways, a blueprint, and regional unity. The notion of claimant states in a sense, capacity is identified through the UNCLOS so the overriding geopolitical context in terms of the legal framework is the Law of the Sea Convention. Within the South China Sea, as was shown by John McManus, there are exclusive economic zones. The claimant states have EEZs. Beyond EEZs, presumably, are the high seas. In the context of the South China Sea as well as in the Arctic where I work predominantly, in a sense, the high seas is an area that is explicitly under international law beyond sovereign jurisdictions. The high seas lend itself to thinking in terms of ways that build common interests. In a sense, the high seas is the international community looking outward from their common interests as opposed to the EEZs and claimant territorial seas looking inward in terms of their national interests. If you look at the region of a whole, in a sense just like the Earth, the region itself has areas that are explicitly under international law beyond sovereign jurisdictions and within sovereign jurisdictions. Seeking that balance or common ground may help to create the type of blueprint that you anticipate.

1:10:26

Mr. James Borton: Very good, thank you Dr. Berkman. The other question I had was, is the establishment, or what I call public-private multilateral partnerships in the South China Sea, marine science surveys to measure and monitor the status, specifically coral reef health—I don't know where we are in terms of those public-private multilateral partnerships. Maybe someone would like to address that?

1:11:06

Dr. John McManus: I can just say I'm picturing a private group getting involved here. There is actually a record of a group hiring a yacht from Singapore many years ago and when they got into the Spratlys, Vietnam sank them. So, there is no private operation going through here other than some of the corporations involved in the drilling in collaboration with various countries. So, it's not really a private thing.

I wanted to mention that Paul Berkman was talking about the EEZs. I showed the same picture that shows the EEZs, that's just on paper. In reality, people can hardly get out there, into their own EEZs because the Chinese claim is very close to the coastlines throughout the area and China clearly dominates the area—most countries with the exception of Taiwan—are concerned with putting any boats in the water out of concern that they'll be rammed or sunk. So that is the reality now and when I went to the Philippine base in 2016, I know that that had been—there was actually a marine lab that had been set up by the University of Philippines. I found out later that the University of the Philippines had not been there since the year 2000 because of this problem. So, in addition to this, we have the problem, as Paul knows and Lina I'm sure, is that you cannot legally do research in another country without permission from that country. Asking for permission is an act of recognizing someone else's sovereignty so no one wants to ask permission. That does not keep us from sharing the votes as Allen was pointing out in James's picture. There is no reason why, in these expeditions China and Taiwan are doing, that they don't invite the other people. That's actually something that happened a bit with the Philippine and Vietnamese cruises in the late 90s and early 2000s. Actually, Liana was on one—China would not allow me on there because I'm clearly an American, but Liana got on board and she actually cruised.

1:14:15

Mr. James Borton: Very good. Thank you, John. I'm now beginning to pick up a few questions and I'd like to share them with all the panelists. This is a question from Greg Long from Radio Free Asia. His question is—for the first set of panelists, or whoever would like to take it—China's distant water fishing fleet and the survey fleet in the South China Sea is ubiquitous and tends to be tied to its foreign policy and armed forces in that area. Is there any indication that China would be sincere about creating and participating in a regional fishery management organization in the South China Sea?

1:15:04

Dr. John McManus: Yes, this is something I deal with all the time. There's this thing that China only deals bilaterally and that does come up when they talk about oil exploration. I looked around and it turns out China is a part of 17 or more land, environment, transboundary agreements. Otherwise, they want to protect the environment and one of those involve four countries. They have a history of doing multilateral agreements. China is involved in several international fishery organizations as well. So, it's not out of the picture that China could do this, it's just that with any country, it kind of depends on who you talk to and who happens to be making the decisions. But I do think it's possible. China does have relationships all over, but all the scientists across SE Asia and China tend to agree—they're very distressed about the damage to the coral reefs, they're worried about the fisheries collapsing, they don't have enough of the data sharing to know for sure what the status of some of these things are and certainly not enough to monitor so we can protect the spots. So, all of the scientists know this as well there are lots and lots of conservationists in China and they tend to be very brave individuals because sometimes you have to be careful of what you say. So, there is a lot of hope and we have to keep getting this out there, but we must fight this with the party line which is, "Everyone knows the South China Sea has belonged to China since time immemorial." That is an official line you get from, I'm sure you've heard this, from a political scientist representative from China. We have to fight that; we have to get the average person in China to realize that's wrong. It's difficult.

I even show a picture with Taiwan as a separate country and some people from China including some students from here see that and say "Why? That's not a separate country." And yet it is, obviously, as Allen would be happy to tell you, so we have to get the information inside China.

There are some other myths—many other people believe that an island belongs to you if it's in your EEZ. That's rampant in the Philippines and Vietnam. This is wrong. Just because it's in your EEZ does not make it yours if it's above high tide. So, we have to kill these myths.

1:18:24

Mr. James Borton: Thank you, John. In fact, to add to that, I have learned that, as you said, several Chinese marine scientists have been participating in some of these regional workshops and there are an increasing number of marine scientists from China who are interested in science cooperation. I have established email communications with them. Again, as you've said, those are brave individuals.

I've just received another question here and I'm going to read this. Science diplomacy may get stymied by geopolitical problems, just as we were alluding to, as is happening today in the region. But can it not be representative and inclusive in nature at the subnational levels in this region—in terms of knowledge, creation, ownership, and resource management. There may be a tendency to de-politicize the debate at times to secure solutions that may leave out many and lead to further conflicts. What are your views on that, for any of the panelists?

1:19:36

Dr. Paul Berkman: I'll jump into that. Geopolitical stymying science diplomacy. This is the problem. It is short-term thinking. At the short timescale of geopolitical dynamics, unless you're a superpower or have access to resources at a global scale, we're all at a disadvantage. However, leveling the playing field happens over time and the further you look into the future, the more equitable the discussion becomes. The reality is that children and young adults that are alive today will be alive in the 22nd century. So, when we talk about across generations, it's not fanciful it's across the 21st century and to a significant extent, there is opportunity to turn the dynamic of the discussion in which China itself has opened the door—which is thinking across the 21st century with the One Belt One Road Initiative. So, in a sense, operating at a geopolitical timescale is a disadvantage. The reality is that we have the opportunity to think across generations. We have tools like the sustainable development goals which think at local to global scales across generations. So, I would note, in terms of the stymying, it's only because it's short-term thinking. Short-term thinking is paralysis. It leads to the gloom and the doom and the nationalism that we face on a regular basis.

Just following up briefly, and then I'll stop, James. The comments that John was making a moment ago about China and bilateral and multilateral, there may be an analogy to explore in the South China Sea and that is the Arctic. I hadn't thought of it until this discussion. But the Arctic states all have sovereign interests and EEZs and the center of the arctic is the high seas—the Central Arctic Ocean high seas. There is a fisheries agreement—the Central Arctic Ocean High Seas Fisheries Agreement—which has, as a basis, a precautionary approach. It may be helpful to think of something analogous to a Central South China High Seas Fisheries Agreement where the participants are not limited in terms of their presence and capacity within EEZs but operate under international law in an area that is clearly beyond sovereign jurisdiction. So, I just introduce that as a potential way of framing questions that may help to diffuse and open doors for dialogues that are otherwise complicated.

1:22:41

Mr. James Borton: Well I'm very happy to hear that, in fact, it seems to me that the Arctic Council might serve as a model for the realization of the South China Sea marine science council and perhaps some of our other panelists would even like to talk about what their experience has been in the formulation in this marine science council and how that might be effectively ramped up as we approach this ecological crisis in the South China Sea.

1:23:22

Dr. Ma Carmen Ablan Lagman: James, I just want to put in a short interjection, putting together what you were saying earlier about private partnerships and this discussion on long-term timelines as opposed to generational, you know generational timelines. I have lived long enough to see the start of this discussion when I was still working with John McManus as an early graduate student and I am now a professor at a university and I kind of think what has been said and what needs to be done has already been articulated so many times again and I will be here even another ten more years I guess. Unless people can find an interest to sustain initiatives outside of their writing of grants and getting funding—that would either come from private partnerships or national governments to be able to sustain the initiatives that will keep the scientific operation going. The other thing is to find creative ways—I think when benefits are being talked about or scaling up small things that we do with initial pump prime funding that would then require the buy-in and resources that are not in the traditional pipelines we have right now. That's all I have to say right now.

1:24:47

Mr. James Borton: Thank you, thank you very much. You know it seems we are right on schedule here in terms of looking at our clock and approaching 90 minutes. I want to just take a moment to reflect on the comments that have been presented by you, distinguished panelists. I want to again thank the East West Center for your generous support and participation in this program. I think as Dr. Berkman just highlighted, perhaps indeed we can take the conversations from this webinar and indeed continue them. I am just delighted that we were able to assemble all of you from different time zones and I generously thank you for your participation and welcome a follow-up to our conversation. And perhaps even the East West Center might have a part 2 of this program down the road. So that's it from me. I don't know if anyone else would like to make any remarks, but I think we're ready—to those of you in Asia I would have to say good night to you—it is late there and thank you again.

1:26:12

Dr. Satu Limaye: James, can I just say from the East West Center's perspective, our mission is, as assigned by the Congress, our mandate is to bring Americans and Asians together on issues of common concern, common challenge, common opportunity, and so I can't think of a topic and a group of folks representing not every country or every issue in the region but certainly an important element set of issues together. I thank you James for collaborating with us on this and I thank all of the participants who spoke today but also all of you online—almost 150 signed up we had I think at one point about close to 90 if I observed correctly—joining us at all kinds of times from all kinds of places so thank you all be safe, be healthy, be happy and we look forward to continued engagement and meeting our mission. Thank you so much indeed.

Appendix C

East-West Lecture—Part 2—Transcript
James Borton

Dr. Ma Carmen Ablan Lagman–Professor at De La Salle University in the Philippines

Speaking from the Philippines, I am giving you a summary of things that are written in a document, and I can give you a copy later. The United Nations Convention on the Law of the Sea may be the single most important regional policy instrument that has changed the face of fisheries in the region and again in the whole world. With this policy, we actually had the ability to call a certain part of the ocean our own, which brings with it the ability to put in more investment because of knowing that we will have a return on those investments.

Just looking at the fisheries' data on the marine fisheries production, South China Sea fishing states, you can see a very sharp increase in the catch value simply because of the investment during that time. But with the rights come responsibilities. Article 63 (1) states, "Where the same stock or stocks of associated species occur within the exclusive economic zones of two or more coastal states, these states shall seek, either directly or through appropriate subregional or regional organizations, to agree upon the measures necessary to coordinate and ensure the conservation and development of such stocks without prejudice to the other provisions of this Part."

If you look at the lists of top countries that are producing or have the highest marine capture fisheries production. Countries like China, Indonesia, Vietnam, the Philippines, Malaysia, and even Thailand, Myanmar, and Taiwan out here are in that really big populus. And that has not changed.

But if you would examine the culture and diet of the people in this region, you can see from Australia, Brazil, Germany, UK, and the United States, just the amount of fish protein in our diet would be almost three times that of our counterparts in the U.S. Therefore, fish plays a very big part in the diet and the importance of culture and livelihood of this region.

Now we are talking about a region in the middle of common territories which has all this information—190 million people live in the coastal areas, catching 17 million tons of fish worth 22 billion dollars in value are landed annually. So this is just to show how important, economically, this region. And aside from the use of fisheries, there is cargo and all the other kinds of uses you have for this region.

Just to give a background of fisheries, if you look at the fish on the left, these are fish that they get from offshore or further from the shore, and the fish on the right you will get closer to the shore or where habitat exists. And if you look at the fishers who are actually catching fish, those with the smaller boats and those with the near short fish are economically at a greater disadvantage at least for the ownership of the protein sources than those catching fish from offshore, further from 15 kilometers or two miles from the shoreline.

So, officially, policies are definitely structured across resources, and resources are structured across habitats. It is not true that the fish that are offshore will just stay offshore, and those inshore will stay inshore, but these habitats, which are very close to the shore, play a very big part in the production of the fish that people catch on both sides of that line. And you are talking about nigrils, egrasses, coral reefs, and all other structures around that area. Of course, as you have seen from Dr. McManus' talk earlier, the central part of the South China Sea has a lot of these areas and habitats existing across it.

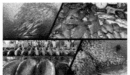

Here are just some examples of habitats that have been rather pristine and beautiful in their current condition. The juvenile fish nursed in this habitat in the protections they afford. In the picture below, here is the image before I went to Vietnam and then again when I came back to the Philippines. You can practically see where the reef used to be. This is just like over twenty years from the time when I first got my license and then coming back. This is not unprecedented in the habitat destruction that Dr. McManus was talking about.

Looking at the profile of fisheries habitats across China, Indonesia, Malaysia, the Philippines, Taiwan, and Vietnam, it basically just says that the shelf area would have a high percentage of the coral reefs in the world, and they are with a lot of people that will actually cause destruction. And it is very hard to tell people that we are losing fish, and you see this at port. When you see this in the market, it is very hard to explain that we are losing our fish. But what they do not understand is that overexploitation is actually changing the nature of the fish that are being caught. And some emphasize how fishing [unintelligible] and the change in the species is happening. There are differences in policies in different countries. Just the number of fishing zones, meaning the ability to go out and fish and what their policies are regarding who catches where varies between these countries. And if you would kind of think about the conflicting policies on fishing zones and issues of closed seasons, we have been through a very tight spot with the Recto Bank incident where Philippine vessels were rammed by Chinese vessels apparently. But, you know, can you just imagine fishing and that you are being told you cannot fish and somebody else is getting fish, which is getting worse. I am not saying somebody's right or wrong here, but this conflict in policies is what is costing these major difficulties around the area. And you are talking about poor fishers in the outskirts of the country.

A study we did with Dr. McManus, Dr. Chen, and many regional partners, looked into the genetic structure of fish around the area. And what we basically found out was a map that shows fish resources go around in circles, and they basically are not defined by national boundaries or EEZ, but mainly one place is covered by many countries and many countries are responsible for this behavior. This further emphasizes the need for that regional fisheries policies that doesn't necessarily have to be associated with ownership, sovereignty, and rule of law. So, just as a list of recommendations.

This booklet which I wrote and it is on the web if you can find it. We need the establishment of a wider scale transboundary fisheries management policies, which may include the establishment of common no-take zones as well as common understanding of closed seasons. This should not be so difficult because there are existing regional, international instruments. But most of all, this could actually be an opportunity to redefine fisheries policies for the small fishers who are least likely to cause habitat destruction. If you look at the difference on paper on the benefits and the destruction for fisheries—small fisheries and large scale fisheries—there really has to be a rethinking about policies.

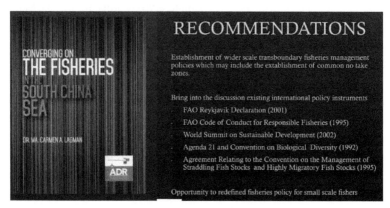

Well, thank you very much Professor Lagman, that was very useful as well. We have had three terrific presentations—the nature of the problems, whether they be environmental on coral reefs, cooperation on fisheries, the nature for the requirements on fish and protein stocks. We now turn to panel two, which looks at the theme of science cooperation, opportunities, tools, and technologies.

Dr. Allen Chen from the Biodiversity Research Center in Taiwan, Dr. Lina Gong, a research fellow with the Rajaratnam School of International Studies in Singapore (RSIS-NTU), Dr. Liana Talaue McManus, a scientist with the GEF-United Nations Environmental Program Marine Plastics Program. Okay, delighted!

Would Dr. Chen please lead us off for seven minutes.

Dr. Allen Chen—Research Fellow at the Department of General Science of Biodiversity Research Center, Academia Sinica, Taiwan.

First of all, I would like to thank the East-West Center for the invitation and the opportunity to see a lot of very good offerings here. It is very delightful to listen to John talk about the resources in the coral reefs in the South China Sea. So, I would like to continue to share my experience working in the region. I hope that my experience shared here will be helpful to continue this talk in this region we all care about. So I will focus on the coral reef interdependence and conservation in the era of changing climate.

Coral reef biodiversity is the hardest in the ocean. The coral reefs occupy less than 300,000 km² and less than 0.2% of the ocean. It is the home for a lot of species associated with coral reefs. It is estimated that there are over 100,000 known species from coral reefs, but millions, including those that are micro-organisms that are still left to discover. There is high diversity that is very important. One billion people rely on healthy coral reefs, including the islands of the Pacific and from the ocean and the Caribbean. It is very high productive in terms of value. It is estimated that nearly 30 billion dollars of global net benefits are provided by the coral reefs, including fisheries and tourists, etc.

The South China Sea is also called the West Philippine Sea or the East Sea, depending upon the country who is talking about and claims sovereignty to this area. But based on the fish species, there are 3,794 fish species are reported here. In terms of Scleractinian species, it has been shown that there are over 571 species, which is more than the qualifying goal region just next to the South China Sea. So this whole area is a very important topic to mention.

There are four major threats to coral reefs. The first one is overfishing, as we have seen from our previous speakers—also, pollution and also habitat destruction. And the most serious are effects caused by climate change due to the increase in seawater temperature. As we are learning from this webinar, these coral reefs are a very important protein provider to our human population. As we have seen in the region, this has already caused big problem issues since protein provides food for the whole South China Sea. In terms of habitat destruction, Singapore found a few years [uncertain]. As John just mentioned, it is a very serious problem. There is a study already shown that increasing artificial island area was inversely proportional to the decreasing of coral reefs. John also published a paper in 2017 that indicated the loss of land and also the coral reef due to the [unsure] by removing the giant clams mainly due to the Chinese fisherman in the South China Sea. But all of this is endangered by the greenhouse gas emissions outside to the air and heat that actually went into the ocean. The increase in the temperature has costs already. This summer, the disruptions have been the most severe near these coral reefs. And also, the impact of climate change has caused increased temperature with mass bleaching of the corals in Taiwan and also seen in the South China Sea.

The IPCC has already published two references in the last year. One is the severity of 1.5°C and the other was published last October about the ocean and cryosphere in a changing climate. All of the scientific evidence is already there.

The coral reef will be impacted by a half degree, but it will be more severe in the ecosystem here. If the temperature goes up two degrees, we will probably lose 99 percent of global coral reefs, including those in the South China Sea, which is very severe. We need to fight this and the greatest attention has to be put into conservation. But if we could just control the temperature, down to 1.5 degrees higher, we will be able to save 10 to 30 percent of the coral reefs on the global scale. So the question is whether will the South China Sea be the 10 to 30 percent. Therefore, there are good science cooperation opportunities in the South China Sea in the era of changing climate. What I would suggest is that there are four major cooperation opportunities. One is the coral reef interdependence, resilience, and governance in the region. For the slide I show here, we have to work with conservation and also not continue to lose the coral reef in the region as was shown in most of the reclaimed islands made by China in the last few years. It has been a great cause of damage. You can see that some island conditions by Taiwan. Conservation will provide hope for the future. The second is oceanic, atmospheric, and biogeochemical processes and typhoons. Also we need to continue get cooperation to do research on oceans in terms of what is happening in the region here. But most importantly, we need to look into the social-ecological system and also food security in the region to provide the region with a better picture. And, we need to call attention to coral reef conservation strategy and maintain under 1.5°C scenario for the future of the coral reefs. One day, I hope we can sail our vessel in the South China Sea and conduct our science projects.

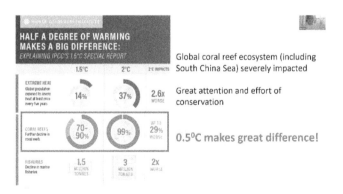

HALF A DEGREE OF WARMING MAKES A BIG DIFFERENCE: EXPLAINING IPCC'S 1.5°C SPECIAL REPORT

Global coral reef ecosystem (including South China Sea) severely impacted

Great attention and effort of conservation

0.5°C makes great difference!

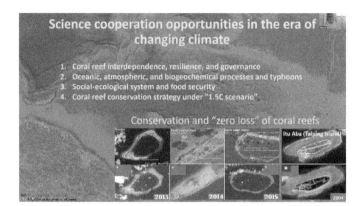

Abbreviations and Acronyms

AEGE	ASEAN Expert Group on Environment
ASEAMS	Association of the Southeast Asian Marine Scientists
ASEAN	Association of Southeast Asian Nations (Brunei, Cambodia, Indonesia, Laos, Malaysia, Myanmar, Philippines, Singapore, Thailand, and Vietnam)
ASEAN Plus Three	10 ASEAN states plus China, Japan, and the Republic of Korea
ASEAN Plus Six	10 ASEAN states plus Australia, China, India, Japan, South Korea, and New Zealand
ASOEN	ASEAN Senior Official on Environment
ASEP	ASEAN Environmental Program
BARMM	Bangsamoro Autonomous Region in Muslim Mindano
BOL	Bangsamoro Organic Law
BRI	Belt and Road Initiative
CCP	Chinese Communist Party

Century of Humiliation	1842–1949, one hundred years during which China was defeated in the Opium Wars, subject to foreign occupation, and decades of internal conflict and civil war, until the founding of the PRC
CFO	Commission on Filipinos Overseas
CGA	Coast Guard Administration
Cha-Cha	A colloquial term for changing the 1987 Filipino Constitution
CNN (CPP)	Communist Party of the Philippines
CNNOC	China National Offshore Oil Corporation
COBSEA	Co-ordinating Body for the Seas of East Asia
DIC	Department of International Cooperation
CoC	Code of Conduct
Covid-19	Coronavirus pandemic that broke out in fall 2019 in Wuhan, China
CPEC	China Pakistan Economic Corridor
CPI	Corruption Perceptions Index
CRS	Congressional Research Service
CSIC	China Shipbuilding Industry Corporation
CSIS	Center for Strategic and International Studies
DFA	Philippines Department of Foreign Affairs
DND	Philippines Department of National Defense
DoC	Declaration of the Conduct of Parties in the South China Sea
EAS/RCU	East Asian Seas Action Plan Regional Coordinating Unit
EDCA	Enhanced Defense Cooperation Agreement
EEC	European Economic Community
EEZ	Exclusive Economic Zone
ESCAP	Economic and Social Commission for Asia and the Pacific
EU	European Union
FAO	Food and Agriculture Organization
FDI	Foreign Direct Investment

Five Eyes Alliance	An intelligence sharing alliance between Australia, Canada, New Zealand, United Kingdom, and the United States founded in 1941
FONOPs	Freedom of Navigation Operations
GDP	Gross Domestic Product
GONGOs	Government-Owned NGOs
GPEA	Greenpeace East Asia
GEF	Global Environment Facility
GIWA	Global International Water Assessment
GPA/LBA	Global Program of Action for the Protection of the Marine Environment from Land-based Activities
Gray Zone Operations	Intense political, economics, informational, and military competition that is just short of war
Great Power	The ability of a sovereign state to exert its influence on a global scale
Hedging	The management of risk relationships between two opposing sides to protect security and sovereignty by diversifying commitments and simultaneously balancing and engaging
Hegemon	A political state having dominant influence or authority over others
ICC	International Criminal Court
ICRAN	International Coral Reef Action Network
ICT	Information and Communication Technology
IMF	International Monetary Fund
IMO	International Maritime Organization
IO	International Organizations
IOC	Intergovernmental Oceanographic Commission
IR	International Relations
IUU	Illegal, unreported, and unregulated fishing practices
JMSU	Joint Marine Seismic Undertaking

KIG	Kalayaan Island Group
MAP	Mediterranean Action Plan
MDT	Mutual Defense Treaty
MFA	Ministry of Foreign Affairs
MNLF	Moro National Liberation Front
Moro	People who live in the southern, predominantly Muslim area of the Philippines
MOU	Memorandum of Understanding
MPA	Marine Protected Area
NATO	North Atlantic Treaty Organization
NGO	Non-governmental Organization
Nine Dash Line	China bases its determination of rights to the South China Sea based upon a nine-dash line drawn upon a map in 1947
NM	Nautical Miles
NOC	National oil companies
NSC	National Security Council
OBOR	One Belt, One Road
OECD	Organization for Economic Cooperation and Development
OMEP	Office for Marine Environmental Protection
ONGC	Indian Oil and Natural Gas Corporation
ONI	Office of Naval Intelligence
Paracel Islands	A disputed archipelago with 130 coral islands and reefs in the South China Sea that is occupied by China and claimed by Taiwan and Vietnam inhabiting a population over 1,000
PCA	Permanent Court of Arbitration
PEMSEA UNEP/ UNDP/IMO Projects	"Prevention and Management of Marine Pollution in the East Asian Seas (1993–1998), and "Building Partnerships for the Environmental Protection and Management of the East Asian Seas (1998–present).
Petro Vietnam	Vietnam Oil and Gas Corporation
PH	The Philippines
PLA	People's Liberation Army (PRC)

PLAAF	People's Liberation Army Air Force (PRC)
PLAN	People's Liberation Army Navy (PRC)
PNOC	Philippine National Oil Company
PRC	People's Republic of China
RP	Responsibility to Protect
RCEP	Regional Comprehensive Economic Partnership
RMPAC	Rim of the Pacific
RMB	Renminbi (Chinese currency)
ROC	Republic of China
SCS	South China Sea
Sinopec	China Petroleum and Chemical Corporation
SOE	State-Owned Enterprise
TAC	Treaty of Amity and Cooperation in Southeast Asia
TPP	Trans-Pacific Partnership
TDA	Transboundary Diagnostic Analysis for the South China Sea
UFWD	United Front Work Department
UN	United Nations
UNCED	United Nations Conference on Environment and Development
UNCLOS	United Nations Convention on the Law of the Sea
UNDP	United Nations Development Programme
UNEP	United Nations Environment Programme
UNEP/DGEF	UNEP Division of Global Environment Facility
UNESCO	UN Educational, Scientific and Cultural Organization
USAID	United States Agency for the International Department
USD	United States Dollar
VAT	Value-Added Tax
VCP	Vietnamese Communist Party
VFA	Visiting Forces Agreement
WHO	World Health Organization
WPS	West Philippine Sea

WTO	World Trade Organization
WWF	World Wide Fund for Nature
ZOPFAN	Zone of Peace, Freedom, and Neutrality

Acknowledgements

This book belongs to many people. Let me start with the fishermen who daily cast their nets into the churning South China Sea. They offered up their stories about declining fish stocks, boats rammed and sometimes even sunk. I know that as I write my book, there's no shortage of stories of fast-moving typhoons capsizing the livelihoods of more fishermen. Men have fished in this ocean since before recorded history. Pollution, over-fishing, storms and encounters with Chinese vessels have diminished their fishing, but still the fishing boats return sometimes heavy in the water, hulls filled with tuna, mackerel, croaker and shrimp.

I am grateful to the Vietnamese boat captains and crew who have generously given me their time and allowed me to board their colorful wooden trawlers. Their honest stories on the South China Sea disputes are genuinely fascinating and arguably, their voices cannot be heard anywhere else but in and around the fishing docks, onboard vessels, in villages, and piers especially found among those hardscrabble lives who shared many details about their lives in Da Nang, Cu Lao Cham and Ly Son Island.

As I was writing this manuscript, a number of published books aided my work with their perspectives and exhaustive research. I am more than obliged to mention, *The South China Sea: The Struggle for Power in Asia*, by Bill Hayton, *Water: Asia's New Battleground* by Brahma Chellaney, *The End of the Line* by Charles Clover, *The Outlaw Ocean* by Ian Urbina, Rachel Carson's *The Sea Among Us, Raging Waters in The South China Sea* by Rachel A. Winston & Ishika Sachdeva, *Wild Sea A History of the Southern Ocean* by Joy McCann, *Navigating Uncertainty in the South China Sea Disputes: An Interdisciplinary Perspective* by Nalanda Roy, *The Brilliant Abyss* by Helen Scales, *Last Days of the Mighty Mekong* by Brian Eyler, and John Steinbeck's *The Log from the Sea of Cortez.*

There's a growing list of colleagues, family, and friends who have offered me their generous advice and guidance along this writing journey: Nena Powell Rice, Travis Simpkins, Piper Walker, Veronica Osborn, Dan Abel, Paul Berkman, Lucio Blanco Pitlo III, Bui Van Nghi, Carla Freeman, Zachary Fillingham, Kent Harrington, Chu Manh Trinh, Nguyen Chu Hoi, Mary Sue Bissell, Brian Eyler, Merrill Pasco, Nguyen Hoang, Tuyen Le Quang, and Nguyen Thanh, Nghi Van Bui, Nguyen Kim Anh, Carl Thayer, Nong Hong, John McManus, Liana McManus, Daniel Pauly, Zev Moses, Lina Gong, Gregory Poling, Keryea Soong, Satu Limaye, Sarah Wang, Khiêm Nguyễn, Huu Thinh, Lap Ngo, Bich T. Tran, Bui Minh Long, La Thang Tung, Kieu Thoan Thu, Kieu-Linh Caroline Valverde, Courtney Weatherby, Meixia Zhao, and Weidong Yu. I also recognize Tay'kiersten Wright, who dutifully assisted me with the bibliography and fact checking.

In the course of my research on the South China Sea, I uncovered many stories of heroism. Luong Thang Dac offered generous details about his family's perilous journey and the risks they faced on the churning sea in their passage to freedom after the Vietnam War.

I owe additional thanks to Hanh Nguyen, who spent countless hours in Vietnamese libraries and also followed up with several fishermen as part of the needed research and translations required to complete this book. Also, I am indebted to Rachel Winston, who completed detailed transcriptions of previous South China Sea conferences and webinars. I am equally grateful to Ao Ava Shen, who generously read early galley proofs, insured placement of proper endnotes and compiled the index.

I also recognize the exceptional talent of Wesley Strickland, a graphic designer and illustrator, who has produced numerous illustrations, and power points for me reflecting environmental South China Sea issues. He is also responsible for this book's impactful cover design.

My appreciation also goes to Le Thu Mach, a Lecturer at Ho Chi Minh National Academy of Politics, who shared valuable insights from her research and doctoral dissertation on social media and green public space.

In addition, I am also grateful to Binh Lai, Deputy Director General at the Diplomatic Academy's East Sea (South China Sea) Institute, since generous invitations were extended to participate in their annual South China Sea programs that benefitted my book's research.

I also single out Nguyen Minh Quang, the co-founder of the Mekong Environment Forum, who has generously helped me to better understand the many challenges of farmers in the Mekong Delta. He was also instrumental in engaging young Vietnamese volunteers from Can Tho University in the development of successful Citizen Science workshops.

I owe as much or more to the Vietnamese people who generously welcomed me into their homes.

Whatever deficiencies either in breadth or depth of subject matter, the sole responsibility for any shortcomings in the final product is all mine.

Endnotes

Chapter One

1. Pham, Charlotte. "Boats in Vietnam." *Encyclopedia of the History of Science, Technology, and Medicine in Non-Western Cultures*, edited by Helaine Selin, Springer, Dordrecht, 2011.
2. Gutierrez, Miren, et al. "China's Distant-Water Fishing Fleet: Scale, Impact and Governance." *ODI*, 1 June 2020, www.odi.org/publications/16958-china-s-distant-water-fishing-fleet-scale-impact-and-governance.
3. "About." *China Dialogue Ocean*, https://chinadialogueocean.net/about/. Accessed 27 Jan. 2021.
4. "The Cattle of Gac Ma: Courage to Defend National Maritime Sovereignty." *BienDong.Net*, 6 Nov. 2018, https://www.southchinasea.com/analysis/1647-the-battle-of-gac-ma-courage-to-defend-national-maritime-sovereignty.html.
5. "New Book Released to Pay Tribute to Truong Sa Martyrs." *Nhan Dan*, 6 July 2018, https://en.nhandan.org.vn/culture/item/6355202-new-book-released-to-pay-tribute-to-truong-sa-martyrs.html.
6. Huang, Cary. "Farewell Comrade: Why Communist China and Vietnam are Drifting Apart." *The South China Morning Post*, 3 Dec. 2017, https://www.scmp.com/week-asia/opinion/article/2122505/farewell-comrade-why-communist-china-and-vietnam-are-drifting.

Chapter Two

1. Pauly, Daniel, and Dirk Zeller. "Catch Reconstructions Reveal that Global Marine Fisheries Catches are Higher than Reported and Declining." *Nature Communications*, vol. 7, no. 10244, 2016, https://www.nature.com/articles/ncomms10244.
2. The interview was completed by telephone between August 21, 2020 and November 21, 2020 with assistance from Hanh Nguyen. Parts of my story were published in Asia Times Online. Borton, James. "Bad Luck Sinks Fisherman's Dreams in South China Sea." *Asia Times*, 15 Dec. 2020, https://asiatimes.com/2020/12/bad-luck-sinks-fishermans-dreams-in-south-china-sea/.
3. Pham, Charlotte Minh Ha. "An Interdisciplinary Approach to the Study of Boats of Central Vietnam." *Journal of Indo-Pacific Archaeology*, vol. 36, Nov. 2016, pp. 25–33, http://dx.doi.org/10.7152/jipa.v36i0.14913.

Chapter Three

1. Pauly, Daniel, et al. "Fisheries, Ecosystems and Biodiversity." *Sea Around Us*, 2016, www.seaaroundus.org/.
2. Nguyen, Thuy Anh. "Science Journals: A New Frontline in the South China Sea Disputes." *Asia Maritime Transparency Initiative*, 15 July 2020, https://amti.csis.org/science-journals-a-new-frontline-in-the-south-china-sea-disputes/.
3. Van Der Kley, Dirk. "The China-Vietnam Standoff: Three Key Factors." *The Interpreter*, 27 Feb. 2017, www.lowyinstitute.org/the-interpreter/china-vietnam-standoff-three-key-factors.
4. Murphey, Martin. "The Abundant Sea: Prospects for Maritime Non-State Violence in the Indian Ocean." *Journal of the Indian Ocean Region*, vol. 8, no. 2, 2012, pp. 173–187, https://doi.org/10.1080/19480881.2012.730751.
5. Munprasit, Aussanee, and Pratakphol Prajakjitt. "Tuna Resource Exploration with Tuna Longline in the South China Sea, Area IV : Vietnamese Waters," *SEAFDEC*, 2001, pp. 29–40, http://hdl.handle.net/20.500.12066/4367.
6. "Fish to 2030: Prospects for Fisheries and Aquaculture." *The World Bank*, no. 83177-GLB, 2013, http://www.fao.org/3/i3640e/i3640e.pdf.
7. Emmers, Ralf. "Indonesia's Role in ASEAN: A Case of Incomplete and Sectorial Leadership." *The Pacific Review*, vol. 27, no. 4, 6 June 2014, pp. 543–562, https://doi.org/10.1080/09512748.2014.924230.
8. Dupont, Alan, and Christopher G Baker. "East Asia's Maritime Disputes: Fishing in Troubled Waters." *The Washington Quarterly*, vol. 37, no. 1, 12 Mar. 2014, pp. 79–98, https://doi.org/10.1080/0163660X.2014.893174.

Chapter Four

1. This chapter is an expanded and revised version of a book review first published in *Asia Sentinel* July 14, 2019 and is kindly reprinted with permission.

2. Vo, Nghia M. *The Vietnamese Boat People 1954 and 1975–1992.* Jefferson, McFarland, 2005.

3. Parsons, Christopher, and Pierre-Louis Vézina. "Migrant Networks and Trade: The Vietnamese Boat People as a Natural Experiment." *IZA*, no. 10112, 2016, http://ftp.iza.org/dp10112.pdf.

4. Papanicolopulu, Irini. "The Duty to Rescue at Sea, in Peacetime and in War: A General Overview." *International Review of the Red Cross*, vol. 98, no. 2, 2016, pp. 491–514, https://library.icrc.org/library/docs/DOC/irrc-902-papanicolopulu.pdf.

5. Ibid.

6. Hsu, Leo. "Praising James Agee." *Foto8*, 10 Dec. 2009, www.foto8.com/live/praising-james-agee/.

7. Sheko, Tania. "Warsan Shire: 'Home.'" *Poem of the Day*, 12 Sept. 2017, http://medium.com/poem-of-the-day/warsan-shire-home-46630fcc90ab.

8. Nash, Stephen E. "The Sound and Fury of the Huey Helicopter." *SAPIENS*, 21 Mar. 2017, www.sapiens.org/column/curiosities/the-sound-and-fury-of-the-huey-helicopter/.

9. Pham, Nguc Luy. *The Freedom Voyage of the Truong Xuan.* Translated by Phan Dam and Thien Vo-Dai, *Hearts of Freedom*, https://heartsoffreedom.org/wp-content/uploads/2019/09/The-Freedom-Voyage-of-the-Truong-Xuan.pdf.

10. Prince Khan, Sadruddin Aga. "Statement of Prince Sadruddin Aga Khan, United Nations High Commissioner for Refugees, to the Third Committee of the United Nations General Assembly, 17 November 1975." *UNHCR*, accessed 8 Aug. 2021, https://www.unhcr.org/en-us/admin/hcspeeches/3ae68fd138/statement-prince-sadruddin-aga-khan-united-nations-high-commissioner-refugees.html.

11. Newland, Kathleen. "Troubled Waters: Rescue of Ayslum Seekers and Refugees at Sea." *Migration Policy Institute*, 3 Jan. 2003, https://www.migrationpolicy.org/article/troubled-waters-rescue-asylum-seekers-and-refugees-sea.

12. The principal maritime treaties that contain rescue obligations include: International Convention on Salvage, April 28, 1899, S. Treaty Doc. No. 102-12 (1991), 1953 U.N.T.S. 193 hereinafter UNCLOS (superseding the four 1958 Law of the Sea conventions: Convention on the Territorial Sea and the Contiguous Zone, April 29, 1958, 15 U.S.T. 1606, 516 U.N.T.S. 82; Convention on Fishing and Conservation of the Living Resources of the High Seas, April 29, 1958, 17 U.S.T. 138, 599 U.N.T.S. 285). For a thorough overview of maritime treaty sources of the rescue obligations, see chapter 8 of *Migration & International Legal Norms*, edited by T. Alexander Aleinikoff, and Vincent Chetail, 2003.

13. Notwithstanding that it was largely superseded by UNCLOS in 1982, the Preamble to the 1958 Convention on the High Seas states that its provisions are "generally declaratory of established principles of international law," Convention on the High Seas, supra note 62, pmbl.; see Brownlie, Ian. *Principles of Public International Law*, 6th ed., Oxford, Oxford University Press, 2003.

14. Excerpts from *Doi Moi Magazine*, July 1975.

Chapter Five

1. "Tourism Industry." *Geopolitical Monitor*, accessed 8 Aug. 2021, https://www.geopoliticalmonitor.com/tag/tourism-industry/.

2. Bobić, Marinko, and Scott N. Romaniuk. "A Fine Balance: China's Need for Resources and Stability in the South China Sea." *Geopolitical Monitor*, 7 July 2015, https://www.geopoliticalmonitor.com/a-fine-balance-chinas-need-for-resources-and-stability-in-the-south-china-sea/.

3. Ibid.

4. Ibid.

5. "Tourism Industry." *Geopolitical Monitor*, accessed 8 Aug. 2021, https://www.geopoliticalmonitor.com/tag/tourism-industry/.

6. Li, Songhai, et al. "Cetaceans under Threat in South China Sea." *Science*, vol. 368, no. 6495, 5 Jun., 2020, pp. 1074–1075, 10.1126/science.abc7557.

7. Lantz, Sandra. *Whale Worship in Vietnam*. 1st ed., Uppsala, Swedish Science Press, 2009.

8. Scarff, James E. "The International Management of Whales, Dolphins, and Porpoises: An Interdisciplinary Assessment." *Ecology Law Quarterly*, vol. 6, no. 2, 1977, pp. 323–427, *JSTOR*, www.jstor.org/stable/24112230. Accessed 10 Jan. 2021.

9. Babcock, Hope M. "Putting a Price on Whales to Save Them: What Do Morals Have to Do with It?" *Environmental Law*, vol. 43, no. 1, 2013, pp. 1–33, *JSTOR*, www.jstor.org/stable/43267663. Accessed 10 Jan. 2021.

10. Lantz, Sandra. *Whale Worship*, p. 27.

11. Tran, Quoc Vuong. *Vietnamese Culture: Study and Interpretations*, Hanoi, 2000, p. 416.

12. *Vietnamese and the Sea*, edited by Nguyen Van Kim. Hanoi, 2011.

13. Ibid., p. 42.

14. Andaya, Barbara Watson. "Seas, Oceans and Cosmologies in Southeast Asia." *Journal of Southeast Asia Studies*, vol. 48, no. 3, 2017, pp. 349–371, 10.1017/S0022463417000534.

15. Earle, Sylvia A. *The World is Blue: How Our Fate and the Oceans are One*. National Geographic Books, 2009.

Chapter Six

1. Quang, Nguyen Minh, and Dao Ngoc Canh. "Saving Vietnam's Floating Markets." *The Diplomat*, 21 Mar. 2017, https://thediplomat.com/2017/03/saving-vietnams-floating-markets/.

2. Lovgren, Stefan. "Southeast Asia's Most Critical River is Entering Uncharted Waters." *National Geographic*, 31 Jan. 2020, https://www.nationalgeographic.com/science/article/southeast-asia-most-critical-river-enters-uncharted-waters#:~:text=Originating%20in%20the%20icy%20headwaters,into%20the%20South%20China%20Sea.

3. Quang, Nguyen Minh. "Is Vietnam in for Another Devastating Drought?" *The Diplomat*, 8 Feb. 2017, https://thediplomat.com/2017/02/is-vietnam-in-for-another-devastating-drought/.

4. Huu, Khoa. "Mekong Delta Hit by Worst Drought Ever." *VNExpress*, 21 Mar. 2020, https://e.vnexpress.net/photo/news/mekong-delta-hit-by-worst-drought-ever-4071241. html#:~:text=In%202016%20it%20lost%208%2C000,12%20provinces%20and%20a%20.city.

5. Beech, Hannah. "China Limited the Mekong's Flow. Other Countries Suffered a Drought." *The New York Times*, 14 Apr. 2020, https://www.nytimes.com/2020/04/13/world/asia/china-mekong-drought.html.

6. "Vietnam, World Bank Sign $560 million to Support Mekong Delta Urban Development and Climate Resilience." *The World Bank*, 11 July 2016, https://www.worldbank.org/en/news/press-release/2016/07/11/vietnam-world-bank-sign-560-million-to-support-mekong-delta-urban-development-and-climate-resilience.

7. Lovgren, Stefan. "Southeast Asia's Most Critical River is Entering Uncharted Waters." *National Geographic*, 31 Jan. 2020, https://www.nationalgeographic.com/science/article/southeast-asia-most-critical-river-enters-uncharted-waters#:~:text=Originating%20in%20the%20icy%20headwaters, into%20the%20South%20China%20Sea.

8. Eyler, Brian, and Courtney Weatherby. "New Evidence: How China Turned Off the Tap on the Mekong River." *Stimson Center*, 13 Apr. 2020, https://www.stimson.org/2020/new-evidence-how-china-turned-off-the-mekong-tap/.

9. Quang, Nguyen Minh, and Dao Ngoc Canh. "Saving Vietnam's Floating Markets." *The Diplomat*, 21 Mar. 2017, https://thediplomat.com/2017/03/saving-vietnams-floating-markets/.

10. "Impacts of Mekong Hydropower Dams and Climate Change in the Lower Mekong Delta." Mekong Environment Forum, 27 Apr. 2020. Webinar.

11. Roberts, Tyson R. "Just Another Dammed River? Negative Impacts of Pak Mun Dam on Fishes of the Mekong Basin." *Natural History Bulletin of the Siam Society*, vol. 41, 1993, pp. 105–133.

12. "Major Paper Plant Halted due to Environmental Concerns." *VietnamNet*, 21 Nov. 2016, http://english.vietnamnet.vn/fms/environment/167335/major-paper-plant-halted-due-to-environmental-concerns.html.

13. "In Southern Vietnam, Hong Kong Paper Mill Renews Concern as Trial Run Nears." *Tuoi Tre News*, 24 Jun. 2016, https://tuoitrenews.vn/news/business/20160624/hong-kong-paper-mill-gives-rise-to-pollution-fears-in-southern-vietnam/22723.html.

Chapter Seven

1. Wright, Lawrence. "The Plague Year." *The New Yorker*, Dec. 2020, www.newyorker.com/magazine/2021/01/04/the-plague-year.

2. Ratner, Ely. "Exposing China's Actions in the South China Sea." *Council on Foreign Relations*, 6 Apr. 2018, www.cfr.org/report/exposing-chinas-actions-south-china-sea.

3. Pauly, Daniel. "Aquacalypse Now." *The New Republic*, 27 Sep. 2009, https://newrepublic.com/article/69712/aquacalypse-now.

4. Sumaila, U. Rashid, et al. "Global Fisheries Subsidies: An Updated Estimate." *Marine Policy*, vol. 69, 2016, pp. 189–193, https://doi.org/10.1016/j.marpol.2015.12.026.

5. Wu, Shicun. "Why US Continues to Stir up South China Sea despite Covid-19 Body Blow." *The South China Morning Post*, 9 May 2020, www.scmp.com/comment/opinion/article/3083194/why-us-continues-stir-south-china-sea-despite-covid-19-body-blow.

Chapter Eight

1. Liu, Zhen. "What's China's 'Nine-Dash Line' and Why Has It Created So Much Tension in the South China Sea." *The South China Morning Post*, 12 July 2016, https://www.scmp.com/news/china/diplomacy-defence/article/1988596/whats-chinas-nine-dash-line-and-why-has-it-created-so.

2. Hayton, Bill. *The South China Sea: The Struggle for Power in Asia*. New Haven, Yale University Press, 2014.

3. Soong, Keryea. Oceanography Department National Sun Yat-Sen University, 17 Feb. 2021.

4. Goron, Coraline. "Ecological Civilization and the Political Limits of a Chinese Concept of Sustainability." *Chinese Perspectives*, vol. 4, 2018, pp. 39–52, http://journals.openedition.org/chinaperspectives/8463.

5. Ibid.

6. Harbone, Alastaire R., et al. "Multiple Stressors and the Functioning of Coral Reefs." *Annual Review of Marine Science*, vol. 9, 2017, pp. 445–468, https://doi.org/10.1146/annurev-marine-010816-060551.

7. Qiu, Jane. "South China Sea: the Gateway to China's Deep-sea Ambitions." *National Science Review*, vol. 4, 2017, pp. 658–663, 10.1093/nsr/nwx107.

8. Li, Yunzhou, et al. "Rethinking Marine Conservation Strategies to Minimize Socio-economic Costs in a Dynamic Perspective." *Biological Conservation*, vol. 224, no. 108512, 2020, 10.1016/j.biocon.2020.108512.

9. "Medicine from the Sea." *Biological Magnetic Resonance Bank*, 7 Oct. 2020, bmrb.io/featuredSys/aspergillic_acid/.

10. Hernández-Fernández, Leslie, et al. "Distribution and Status of Living Colonies of *Acropora* spp. in the Reef Crests of a Protected Marine Area of the Caribbean (Jardines de la Reina National Park, Cuba)." *PeerJ*, vol. 7, no. e6470, 21 Feb. 2019, doi:10.7717/peerj.6470.

11. Hoegh-Guldberg, Ove, et al. "People and the Changing Nature of Coral Reefs." *Regional Studies in Marine Science*, vol. 30, 2019, https://doi.org/10.1016/j.rsma.2019.100699.

12. Bruno, John F., and Elizabeth R. Selig. "Regional Decline of Coral Cover in the Indo-Pacific: Timing, Extent, and Subregional Comparisons." *PLOS One*, vol. 2, 2007, https://journals.plos.org/plosone/article?id=10.1371/journal.pone.0000711.

Chapter Nine

1. Williams, Robert D. "Tribunal Issues Landmark Ruling in South China Sea Arbitration." *Lawfare*, 12 July 2016, www.lawfareblog.com/tribunal-issues-landmark-ruling-south-china-sea-arbitration.

2. "United Nations Convention on the Law of the Sea." *The United Nations*, 10 Dec. 1982, https://www.un.org/depts/los/convention_agreements/texts/unclos/unclos_e.pdf.

3. Diep, Vo Ngoc. "Vietnam's Note Verbale on the South China Sea." *Asia Maritime Transparency Initiative*, 5 May 2020, amti.csis.org/vietnams-note-verbale-on-the-south-china-sea/.

4. Salako, Solomon. "Entitlement to Islands, Rocks and Low – Tide Elevations in the South China Sea: Geoeconomics versus Rule of Law." *International Law Research*, vol. 7, no. 1, 2018, 10.5539/ilr.v7n1p247.

5. Dao, Toan. "Vietnam Sets Goal of USD 10.5 Billion in Seafood Exports for 2019." *SeafoodSource Official Media*, 15 Feb. 2019, www.seafoodsource.com/news/supply-trade/vietnam-sets-goal-of-usd-10-5-billion-in-seafood-exports-in-2019.

6. Godfrey, Mark. "New Data Indicates Big Jump in China Distant-Water Catch." *SeafoodSource Official Media*, 18 Aug. 2020, www.seafoodsource.com/news/supply-trade/new-data-indicates-big-jump-in-china-distant-water-catch.

7. Thanh, Tran Xuan. "Vietnam Fisheries Resource Surveillance Force Stands Side by Side with Fishermen in Exploiting Fish and Seafood and Protecting Sea and Island" *TCQPTD National Defense Journal*, 19 Jan. 2017, tapchiqptd.vn/en/theory-and-practice/vietnam-fisheries-resource-surveillance-force-stands-side-by-side-with-fishermen-in-exploi/9719.html.

8. Thushari, GGN, and JDM Senevirathna, "Plastic Pollution in the Marine Environment." *Heliyon*, vol. 6, no. 8, 27 Aug. 2020, https://doi.org/10.1016/j.heliyon.2020.e04709.

Chapter Ten

1. "Environment and Law." *World Ocean Review*, vol. 3, 2014, worldoceanreview.com/en/wor-3/environment-and-law/international-commitments/.

2. Storey, Ian, and Cheung-yi Lin. *The South China Sea Dispute: Navigating Diplomatic and Strategic Tensions.* Singapore, ISEAS-Yusof Ishak Institute, 2016.

3. Thomas, Michael. "Fish, Food Security, and Future Conflict Epicenters." *The Centre for Climate and Security*, June 2017, https://climateandsecurity.org/wp-content/uploads/2017/06/10_fish-conflict.pdf.

4. Bale, Rachel. "One of the World's Biggest Fisheries is On the Verge of Collapse." *National Geographic*, 29 Aug. 2016, https://news.nationalgeographic.com/2016/08/wildlife-south-china-sea-overfishing-threatens-collapse/.

5. Kraska, James. "The Lost Dimension: Food Security and the South China Sea Disputes." *Harvard National Security Journal*, 26 Feb. 2015, harvardnsj.org/2015/02/the-lost-dimension-food-security-and-the-south-china-sea-disputes/.

6. Glaser, Bonnie. "Armed Clash in the South China Sea, 14 Contingency Planning Memorandum." *Council on Foreign Relations*, 2012, http://www.cfr.org/east-asia/armed-clash-south-china-sea/p27883. Accessed on 24 Mar. 2013.

7. Ibid, p. 2.

8. Grotius, Hugo. *The Free Sea.* Carmel, Liberty Fund, 2012. muse.jhu.edu/book/17941.

9. The White House Office. "National Strategy for the Arctic Region." 2013, https://obamawhitehouse.archives.gov/sites/default/files/docs/nat_arctic_strategy.pdf.

10. Briner, Jason P., et al. "Rate of Mass Loss from the Greenland Ice Sheet will Exceed Holocene Values This Century." *Nature*, vol. 586, 2020, pp. 70–74, https://doi.org/10.1038/s41586-020-2742-6.

11. Luterbacher, Jürg, et al. "United in Science 2020." *World Meteorological Organization*, 25 Oct. 2020, public.wmo.int/en/resources/united_in_science.

12. Kolcz-Ryan, Marta. "An Arctic Race: How the United States' Failure to Ratify the Law of the Sea Convention Could Adversely Affect Its Interest in the Arctic." *University of Dayton Law Review*, vol. 35, 2009–2010, pp. 149–173.

Chapter Eleven

1. "Blue Dot Network." *US Department of State*, 1 Dec. 2020, www.state.gov/blue-dot-network/.
2. Bhatia, Rajiv. "Assessing the 35th ASEAN Summit." *Gateway House*, 28 Nov. 2019, www.gatewayhouse.in/35-asean-summit/.
3. Montague, Zach. "Can the 'Blue Dot Network' Really Compete with China's Belt and Road?" *World Politics Review*, 4 Dec. 2019, https://www.worldpolitics-review.com/articles/28385/can-the-blue-dot-network-really-compete-with-china-s-belt-and-road.
4. Heathcote, Chris. "Forecasting Infrastructure Investment Needs for 50 Countries, 7 Sectors through 2040." *World Bank Blogs*, 10 Aug. 2017, https://blogs.world-bank.org/ppps/forecasting-infrastructure-investment-needs-50-countries-7-sectors-through-2040.
5. Goodman, Matthew P., et al. "Connecting the Blue Dots." *Center for Strategic & International Studies*, 26 Feb. 2020, https://www.csis.org/analysis/connecting-blue-dots.
6. Chairman of ASEAN. "Statement of the 35th ASEAN Summit Bangkok/Nonthaburi, 3 November 2019 Advancing Partnership for Sustainability." *ASEAN Thailand 2019*, 3 Nov. 2019, https://www.asean2019.go.th/en/news/chair-mans-statement-of-the-35th-asean-summit-bangkok-nonthaburi-3-november-2019-advancing-partnership-for-sustainability/.
7. "The Launch of Multi-Stakeholder Blue Dot Network." *DFC - US International Development Finance Corporation*, 4 Nov. 2019, www.dfc.gov/me dia/opic-press-releases/launch-multi-stakeholder-blue-dot-network.
8. Diop, Makhtar. "Accelerating Vietnam's Path to Prosperity." *World Bank Blogs*, 21 Feb. 2019, blogs.worldbank.org/voices/accelerating-vietnams-path-prosperity.
9. Jones, Lee, and Shahar Hameiri. "Debunking the Myth of 'Debt-Trap Diplomacy.'" *Chatham House*, 19 Aug. 2020, www.chathamhouse.org/2020/08/debunking-myth-debt-trap-diplomacy.
10. Anh, Viet. "Vietnam Still Wary of China's Belt and Road Initiative." *VnExpress International*, 16 Nov. 2018, e.vnexpress.net/news/news/vietnam-still-wary-of-china-s-belt-and-road-initiative-3839937.html.
11. Dodwell, David. "Tillerson's Belt and Road Lament Shows China's Growing Influence." *South China Morning Post*, 23 Mar. 2018, www.scmp.com/business/global-economy/article/2138539/tillersons-final-warning-belt-and-road-financing-only-proves.

12. Chatzky, Andrew, and James McBride. "China's Massive Belt and Road Initiative." *Council on Foreign Relations*, 28 Jan. 2020, www.cfr.org/backgrounder/chinas-massive-belt-and-road-initiative.

Chapter Twelve

1. A version of my article was first published in *China Dialogue Ocean* on July 22, 2019 and it has been republished with kind permission.
2. "Life Below Water." *The World Bank*, datatopics.worldbank.org/sdgatlas/archive/2017/SDG-14-life-below-water.html. Accessed 13 Aug. 2021.
3. "Western & Central Pacific Fisheries Commission (WCPFC)." *Western & Central Pacific Fisheries Commission*, 19 June 2004, www.wcpfc.int/.
4. "The Link Between Effective Fisheries Management and Ending Harmful Subsidies." *The Pew Charitable Trusts*, 5 Apr. 2019, www.pewtrusts.org/en/research-and-analysis/issue-briefs/2019/04/the-link-between-effective-fisheries-management-and-ending-harmful-subsidies.
5. "Five Questions for: James Movick, Director-General, Pacific Islands Forum Fisheries Agency." *Business Advantage PNG*, 20 Feb. 2018, www.businessadvantagepng.com/five-questions-for-james-movick-director-general-pacific-islands-forum-fisheries-agency/.
6. Maefiti, John. *Pacific Islands Tuna Industry Association*, 10 July 2019, pacifictuna.org/.
7. University of British Columbia. "Chinese Foreign Fisheries Catch 12 Times More than Reported, Study Shows." *ScienceDaily*, 3 April 2013, www.sciencedaily.com/releases/2013/04/130403104210.htm.
8. Seidel, H., and Padma N. Lal. *Economic Value of the Pacific Ocean to the Pacific Island Countries and Territories.* Gland, Switzerland, IUCN, 2010. https://www.iucn.org/sites/dev/files/import/downloads/economic_value_of_the_pacific_ocean_to_the_pacific_island_countries_and_territories_p.pdf.
9. US Department of Commerce. *National Oceanic and Atmospheric Administration*, 2021, www.noaa.gov/.
10. "Nauru Agreement." *Pacific Islands Forum Fisheries Agency (FFA)*, 2020, www.ffa.int/nauru_agreement. Accessed 13 Aug. 2021.
11. Parker, Sam, and Gabrielle Chefitz. "Debt Book Diplomacy." *Belfer Center for Science and International Affairs*, 24 May 2018, www.belfercenter.org/publication/debtbook-diplomacy.
12. Zhang, Jian. "China's Role in the Pacific Islands Region." *APCSS*, 2015, https://apcss.org/wp-content/uploads/2015/08/C3-China-Pacific-Zhang.pdf.

13. Reuters. "Payments Due: Pacific Islands in the Red as Debts to People's Republic of China Mount." *Indo-Pacific Defense Forum*, 27 Aug. 2018, ipdefenseforum. com/2018/08/payments-due-pacific-islands-in-the-red-as-debts-to-peoples-republic-of-china-mount/.

Chapter Thirteen

1. Rothwell, Donald R. "Could Law Save the South China Sea from Disaster?" *The National Interest*, 26 July 2016, nationalinterest.org/blog/the-buzz/could-law-save-the-south-china-sea-disaster-17123.
2. Wijaya, Lupita, and Cheryl Pricilla Bensa. "Indonesian Mainstream News Coverage of the South China Sea Disputes: A Comparative Content Analysis." *Asian Politics & Policy*, vol. 9, no. 2, Apr. 2017, pp. 331–336.
3. Maragos, J.E., Crosby, M.P., and J.W. McManus. "Coral Reefs and Biodiversity: A Critical and Threatened Relationship." *Oceanography*, vol. 9, no.1, 1996, pp. 83–99. https://tos.org/oceanography/assets/docs/9-1_maragos.pdf.
4. Ibid., 92.
5. Hughes, Terry P., et al. "Global Warming and Recurrent Mass Bleaching of Corals." *Nature*, vol. 543, no. 7645, 2017, pp. 373–377. https://doi.org/10.1038/nature21707.
6. Wilkinson, C., DeVantier, L., Talaue-McManus, L., Lawrence, D. and D. Souter. *Global International Waters Assessment: South China Sea, GIWA Regional Assessment 54*. Kalmar, University of Kalmar, UN Environment Programme, 2005. www.unenvironment.org/resources/report/global-international-waters-assessment-south-china-sea-giwa-regional-assessment-54.
7. McManus, John, Shao, Kwang-Tsao, and Szu Yin Lin. "Toward Establishing a Spratly Islands International Marine Peace Park: Ecological Importance and Supportive Collaborative Activities with an Emphasis on the Role of Taiwan." *Ocean Development & International Law*, vol. 41, no. 3, pp. 270–280. 10.1080/00908320.2010.499303.
8. Ibid.
9. Zbicz, Dorothy. "Global List of Complexes of Internationally Adjoining Protected Areas." *Trans-boundary Protected Areas for Peace and Co-operation*, edited by Adrian Philipps, IUCN, Gland, Switzerland, 2001, pp. 64–75.
10. Yang, Alan H. "The South China Sea Arbitration and Its Implications for ASEAN Centrality." *Asian Yearbook of International*, vol. 21, 2015, pp. 83–95. https://doi.org/10.1163/9789004344556_006.

Chapter Fourteen

1. Bruun, Ole. "Environmental Protection in the Hands of the State: Authoritarian Environmentalism and Popular Perceptions in Vietnam." *The Journal of Environment & Development*, vol. 29, no. 2, 2020, pp. 171–195. https://doi.org/10.1177%2F1070496520905625.
2. Carson, Rachel. *Silent Spring*. Boston, Houghton Mifflin, 2002.
3. Le, Thu Mach. "The Emergence of Social Media and the Green Public Sphere in Vietnam." *Monash University*, Thesis, 2019. https://doi.org/10.26180/5dd268d62c266.
4. Ibid.
5. Vu, N.A. "Grassroots Environmental Activism in an Authoritarian Context: The Trees Movement in Vietnam." *Voluntas: International Journal of Voluntary and Nonprofit Organizations*, vol. 28, 2017, pp. 1180–1208. https://doi.org/10.1007/s11266-017-9829-1.
6. "Vietnam Internet Statistics 2020." *VNetwork*, 19 Feb. 2020, https://vnetwork.vn/en/news/thong-ke-internet-viet-nam-2020.
7. Le, Quang Binh, et al. *Reports on Movements to Protect 6700 Trees in Hanoi*. Hanoi, Hong Duc, 2015. http://isee.org.vn/Content/Home/Library/472/reports-on-movements-to-protect-6700-trees-in-hanoi..pdf.
8. Louv, Richard. *The Nature Principle: Human Restoration and the End of Nature-Deficit Disorder*. Chapel Hill, Algonquin Books, 2011.
9. Dang, Hung. "Pandend - Cộng đồng kết nối Y khoa phòng chống Covid-19." *Facebook*, 29 Jan. 2020, https://www.facebook.com/groups/155042985943381.
10. *Citizen Science Global Partnership*. http://citizen scienceglobal.org. Accessed 14 Aug. 2021.
11. "Cộng đồng kết nối Y khoa phòng chống Covid-19." *Facebook*, Apr. 4, 2020. https://www.facebook.com/groups/2781917808530654.
12. Le, Quoc Vinh. "Chung Tay Chống Tin Giả - Fake News." *Facebook*, 4 Feb. 2020, https://www.facebook.com/groups/antifakenews.vn.
13. Nguyen, Thao Thi Phuong, et al. "Fake News Affecting the Adherence of National Response Measures During the COVID-19 Lockdown Period: The Experience of Vietnam." *Frontiers in Public Health*, vol. 8, Sept. 2020, p. 589872. *PubMed Central*, https://doi.org/10.3389/fpubh.2020.589872.

Chapter Fifteen

1. DeRidder, Kim J., and Santi Nindang. "Southeast Asia's Fisheries Near Collapse from Overfishing." *The Asia Foundation*, 29 Mar. 2018, asiafoundation.org/2018/03/28/southeast-asias-fisheries-near-collapse-overfishing/.

2. Rosenberg, D. "Fisheries management in the South China Sea." *Security and International Politics in the South China Sea: Towards a Cooperative Management Regime*, edited by S. Bateman and R. Emmers, Routledge, 2009, pp. 61–79.

3. *The Notre Dame Global Adaptation Index ND-GAIN.* U of Notre Dame, 2013, www.gain.idex.org. Accessed 14 Aug. 2021.

4. Cheung, William L., et al. "Signature of Ocean Warming in Global Fisheries Catch." *Nature*, vol. 497, 2013, pp. 365–368. https://www.nature.com/articles/nature12156

5. "UN Food and Agriculture Organization, Committee on Fisheries." *The Pew Charitable Trusts*, 28 Oct. 2014, www.pewtrusts.org/en/research-and-analysis/articles/2014/12/un-food-and-agriculture-organization-committee-on-fisheries.

6. Chalk, Peter. "Illegal Fishing in Southeast Asia: A Multibillion-Dollar Trade with Catastrophic Consequences." *The Strategist*, 14 July 2017, www.aspistrategist.org.au/illegal-fishing-southeast-asia-multibillion-dollar-trade-catastrophic-consequences/.

7. Derrick, Brittany, et al. "Thailand's Missing Marine Fisheries Catch (1950–2014)." *Frontiers*, 28 Nov. 2017, www.frontiersin.org/articles/10.3389/fmars.2017.00402/full.

8. "Part VII High Seas." *United Nations*, www.un.org/Depts/los/convention_agreements/texts/unclos/part7.htm. Accessed 14 Aug. 2021.

9. Agnew, D.J., et al. "Estimating the Worldwide Extent of Illegal Fishing." *PloS One*, vol. 4, no. 2, 2009, 10.1371/_journal._pone.0004570.

10. Brett, Annie, et al. "Ending Illegal Fishing: Data Policy and the Port State Measures Agreement." *World Economic Forum*, 2019, http://www3.weforum.org/docs/WEF_Ending_Illegal_Fishing.pdf.

Chapter Sixteen

1. A version of this chapter was first published in *East Asia Forum Quarterly*, vol. 12, no. 1, 2020, and it has been republished with kind permission.

2. "American Corners." *U.S. Embassy in the Democratic Republic of the Congo*, 13 Feb. 2019, cd.usembassy.gov/education-culture/american-spaces/american-corners/.

3. "About Vietnam—Political System." *Socialist Republic of Vietnam Government Portal*, www.chinhphu.vn/portal/page/portal/English/TheSocialistRepublicOfVietnam/AboutVietnam/AboutVietnamDetail?categoryId=10000103&articleId=10001578. Accessed 14 Aug. 2021.

4. Phan, Hai-Dang. "A Deluge of New Vietnamese Poetry." *Asymptote*, www.asymptotejournal.com/special-feature/haidang-phan-on-contemporary-vietnamese-poets/. Accessed 14 Aug. 2021.

5. Kimball, Jeffery. Review of *Road to Disaster: A New History of America's Descent into Vietnam*, by Brian VanDeMark. *H-Net Reviews*, Jan. 2019, http://www.h-net.org/reviews/showrev.php?id=53340.

6. Thi, Hue Hoang, and HaHoang Thi Hong. "Acculturation in Vietnamese Contemporary Literature." *International Journal of Communication and Media Studies (IJCMS)*, vol. 6, no. 3, June 2016, https://papers.ssrn.com/sol3/papers.cfm?abstract_id=2835456.

7. Le, Minh Khue. *The Distant Stars*, WordPress, 2016. *Zine Library*, https://zinelibrary.files.wordpress.com/2016/02/distantstars.pdf.

8. "Temple of Literature, Hanoi." *Wikipedia*, 6 Feb. 2021, en.wikipedia.org/wiki/Temple_of_Literature,_Hanoi.

Chapter Seventeen

1. Nguyễn, Xuân Phúc. "Chairman's Statement of the 23rd ASEAN-China Summit." *ASEAN*, 20 Nov. 2020, https://asean.org/storage/47-Final-Chairmans-Statement-of-the-23rd-ASEAN-China-Summit.pdf.

2. Earp, Hannah S., and Arianna Liconti. "Science for the Future: The Use of Citizen Science in Marine Research and Conservation." *YOUMARES 9 - The Oceans: Our Research, Our Future: Proceedings of the 2018 Conference for YOUng MArine RESearcher in Oldenburg, Germany*, edited by Simon Jungblut, et al., Springer International Publishing, 2020, pp. 1–19, https://doi.org/10.1007/978-3-030-20389-4_1.

3. Appeltans, Ward, et al. "Ocean Biogeographic Information System (OBIS)." *Census of Marine Life*, 2010, www.coml.org/global-marine-life-database-obis/.

4. Pendleton, Linwood, et al. "Opinion: We Need a Global Movement to Transform Ocean Science for a Better World." *Proceedings of the National Academy of Sciences*, vol. 117, no. 18, May 2020, pp. 9652–9655, https://doi.org/10.1073/pnas.2005485117.

5. Howe, B. M., and T. McGinnis. "Sensor Networks for Cabled Ocean Observatories." *Proceedings of the 2004 International Symposium on Underwater Technology*, 2004, pp. 113–20. *IEEE Xplore*, https://doi.org/10.1109/UT.2004.1405499.

6. Louv, Richard, and John W. Fitzpatrick. *Citizen Science: Public Participation in Environmental Research*, edited by Janis L. Dickinson and Rick Bonney, 1st ed., Cornell University Press, 2012. *JSTOR*, www.jstor.org/stable/10.7591/j.ctt7v7pp.

7. Conrad, C., and Krista G. Hilchey. "A Review of Citizen Science and Community-Based Environmental Monitoring: Issues and Opportunities." *Environmental Monitoring and Assessment*, 2011. *Semantic Scholar*, https://doi.org/10.1007/s10661-010-1582-5.

8. Fulton, Bill, et al. "Microplastics." *Living Ocean*, www.livingocean.org.au/microplastics.html.

9. Leape, Jim., Mark Abbott, Hide Sakaguchi, et al. *Technology, Data and New Models for Sustainably Managing Ocean Resources.* Washington, DC, World Resources Institute, 2020. https://oceanpanel.org/sites/default/files/2020-01/19_HLP_BP6_V4.pdf.

10. Tjossem, Sara. *Fostering Internationalism Through Marine Science: The Journey with PICES.* Basel, Springer, 2018.

Chapter Eighteen

1. Kindly note that parts of this chapter were first published in *Strategic Vision*, vol. 5, no. 29, October 2016.

2. Teh, Louise S. L., et al. "What Is at Stake? Status and Threats to South China Sea Marine Fisheries." *Ambio*, vol. 46, no. 1, Feb. 2017, pp. 57–72. *Springer Link*, https://doi.org/10.1007/s13280-016-0819-0.

3. Stobutzki, I. C., et al. "Decline of Demersal Coastal Fisheries Resources in Three Developing Asian Countries." *Fisheries Research*, vol. 78, no. 2, May 2006, pp. 130–42. *ScienceDirect*, https://doi.org/10.1016/j.fishres.2006.02.004.

4. Pitcher, Tony, et al. "Not Honouring the Code." *Nature*, vol. 457, no. 7230, Feb. 2009, pp. 658–59. https://doi.org/10.1038/457658a.

5. Borton, James. "Marine Peace Park Plan Offers Promise for South China Sea." *Geopolitical Monitor*, 26 Oct. 2015, https://www.geopoliticalmonitor.com/marine-peace-park-plan-offers-promise-for-south-china-sea/.

6. East West Center webinar, September 30, 2020.

7. Wang, Bin. "The Outlook for the Establishment and Management of Marine Protected Area Network in China." *International Journal of Geoheritage and Parks*, vol. 6, no. 1, June 2018, pp. 32–42. *ScienceDirect*, https://doi.org/10.17149/ijg.j.issn.2210.3382.2018.01.003.

8. Kraft, Herman, et al. "Online Event: Taiwan and Indo-Pacific Regional Security Architecture Conference - Day 2." *Center for Strategic and International Studies*, 13 Jan. 2021, www.csis.org/events/online-event-taiwan-and-indo-pacific-region-al-security-architecture-conference-day-2.

9. Ibid.

10. Hammer, Leonard, and James Borton. "Taiping Island Ideal for a Global Science Peace Park." *Asia Times*, 2 Mar. 2021, https://asiatimes.com/2021/03/taiping-island-ideal-for-a-global-science-peace-park/.

11. Ibid.

12. Dokken, Karin. "Environment, Security and Regionalism in the Asia-Pacific: Is Environmental Security a Useful Concept?" *The Pacific Review*, vol. 14, no. 4, Jan. 2001, pp. 509–30. https://doi.org/10.1080/09512740110087311.

13. "Mischief Reef." *Asia Maritime Transparency Initiative*, amti.csis.org/mischief-reef/.

14. Li, Jianwei, et al. "Closing the Net Against IUU Fishing in the South China Sea: China's Practice and Way Forward." *Journal of International Wildlife Law & Policy*, vol. 18, no. 2, Jul. 2015, pp. 139–164. https://doi.org/10.1080/13880292.2015.1044799.

15. "United Nations Convention on the Law of the Sea." *The United Nations*, 10 Dec. 1982, https://www.un.org/depts/los/convention_agreements/texts/unclos/unclos_e.pdf. Accessed 15 Aug. 2021.

Chapter Nineteen

1. This chapter I authored, "Science Diplomacy and Dispute Management in the South China Sea," was first published 2018 in *Enterprises, Localities, People, and Policy in the South China Sea Beneath the Surface* and is reproduced with permission of Palgrave Macmillan.

2. "History of IIASA." *International Institute for Applied Systems Analysis*, 19 Nov. 2020, http://www.iiasa.ac.at/web/home/about/whatisiiasa/history/history_of_iiasa.html.

3. Ibid.

4. *New Frontiers in Science Diplomacy*. London, Royal Society, 12 Jan. 2010, https://royalsociety.org/topics-policy/publications/2010/new-frontiers-science-diplomacy/. Accessed 15 Aug. 2021.

5. Knoblich, Ruth. "The Role of Science and Technology in the Dynamics of Global Change and the Significance of International Knowledge Cooperation in the Post-Western World: An Interview with Dirk Messner." *The Global Politics of Science and Technology*, vol. 1, edited by Maximilian Mayer, Mariana Carpes, and Ruth Knoblich, Springer, 2014. *ResearchGate*, https://doi.org/10.1007/978-3-642-55007-2_15.

6. Berkman, Paul. *Science Diplomacy*. Washington, D.C., Smithsonian Institution Scholarly Press, 2011.

7. Nguyen, Chu Hoi, "Re: Question for Attribution in Academic Article Science & Diplomacy Journal". Received by James Borton, 1 Oct. 2016.

8. Nguyen, Chu Hoi, ibid.

9. Hong, Nong. "Marine Environmental Security As A Driving Force Of Cooperation In The South China Sea." Taiwan and the 2016 Elections: The Road Ahead, the Twenty-Fourth Annual Conference on Taiwan Affairs, Walker Institute of International and Area Studies University of South Carolina, US, 23–25 Sep. 2016.

10. Page, Jeremy, and Trefor Moss. "South China Sea Ruling Puts Beijing in a Corner." *The Wall Street Journal*, 11 July 2016, http://www.wsj.com/articles/south-china-sea-ruling-puts-beijing-in-a-corner-1468365807.

11. Moss, Trefor. "South China Sea Ruling Could Pose Dilemma for Philippines' Rodrigo Duterte." *The Wall Street Journal*, 12 July 2016, http://www.wsj.com/articles/tribunal-ruling-on-south-china-sea-could-create-dilemma-for-philippines-duterte-1468323025.

12. Bergenas, Johan, and Ariella Knight, "Secure Oceans: Collaborative Policy And Technology Recommendations for the World's Largest Crime Scene." *Stimson Center*, 7 June 2016, https://www.stimson.org/2016/secure-oceans-collaborative-policy-and-technology-recommendations-world-largest-crime/.

13. Ramos, Fidel V. "Breaking the Ice in the South China Sea." *The Strait Times*, 13 Oct. 2016, http://www.straitstimes.com/opinion/breaking-the-ice-in-the-south-china-sea.

14. Gomez, E. D. "Destroyed Reefs, Vanishing Giant Clams: Marine Imperialism." *In the Wake of Arbitration: Papers from the Sixth Annual CSIS South China Sea Conference*, edited by Murray Hiebert, Gregory Poling, and Conor Cronin, Rowman & Littlefield, 2017, pp. 113–123.

15. *The UNEP/GEF South China Sea Project*, United Nations Environment Program, 22 Sep. 2016, http://www.unepscs.org/Project_Background.html.

16. John McManus cited in Borton, James. "After The Hague, ASEAN Must Reach an Ecological Security Consensus." *Geopolitical Monitor*, 19 July 2016, https://www.geopoliticalmonitor.com/after-the-hague-asean-must-reach-an-ecological-security-consensus/.

17. Pitcher, Tony J., et al. "Marine Reserves And the Restoration of Fisheries And Marine Ecosystems in The South China Sea." *Bulletin Of Marine Science*, vol. 66, no. 3, 2000, pp. 543–566.

18. Parameswaran, Prashanth. "America's New Maritime Security Initiative for Southeast Asia." *The Diplomat*, 2 Apr. 2016, http://thediplomat.com/2016/04/americas-new-maritime-security-initiative-for-southeast-asia/.

19. Yap, Helen T., and Josh Brann. "Reversing Environmental Degradation Trends in the South China Sea and Gulf of Thailand – Terminal Evaluation." *United Nations Environment Programme*, 2009, https://wedocs.unep.org/bitstream/handle/20.500.11822/7400/Terminal_evaluation_of_the_UNEP_GEF_project_Reversing_environmental_degradation_trends_in_the_South_China_Sea_and_Gulf_of_Thailand.pdf?sequence=1&isAllowed=y.

20. *Report of the Sixteenth meeting of the Coordinating Body on the Seas of East Asia (COBSEA) on the East Asian Seas Action Plan.* Bangkok, United Nations Environment Program, 2001. https://wedocs.unep.org/bitstream/handle/20.500.11822/29075/CBSA16.pdf?sequence=1&isAllowed=y.

21. Bergenas, Johan, and Ariella Knight.

22. du Rocher, Sophie Boisseau. "Scientific Cooperation in the South China Sea: Another Lever for China." *The Strategist*, 1 Oct. 2015, http://www.aspistrategist.org.au/scientific-cooperation-in-the-south-china-sea-another-lever-for-china/.

23. *New Frontiers in Science Diplomacy.*

24. Bensurto Jr., Henry S. *Cooperation in the South China Sea: Views on the Philippines-Vietnam Cooperation on Maritime and Ocean Concerns.* Second International Workshop on the South China Sea. Hanoi, Thế Giới Publisher, 2011.

25. Chu, Manh Trinh, "Re: Question/Answer for Publication from James Borton." Received by James Borton, 10 Oct. 2016.

26. Susskind, Lawrence. "Re: Follow-up from James Borton Researcher/Writer on Environmental Security in the South China Sea." Received by James Borton, 10 Oct. 2016.

27. Vietnamese not-for-profit and non-governmental organization established in 2008. See more details at: http://cecenter.org.vn/Home.asp?mnz=514&mno=0&languageID=1.

28. An organization aiming at building and improving skills of youth from the Mekong Delta. See more details at mdy.vn.

29. A leading Vietnamese NGO in the field of coastal and marine ecosystem conservation, striving for a coastal zone of Vietnam with healthy ecosystems and a good quality of life for coastal communities. See more details at: http://mcdvietnam.org/.

30. Delizo, Michael Joe T., and Llanesca T. Panti. "Dispute Not Over Yet." *TCA Regional News*, 2016.

31. "Thailand, Indonesia Call for Peace and Stability in South China Sea Ahead of Tribunal Ruling." *The Strait Times*, 12 Jul. 2016, http://www.straitstimes.com/asia/se-asia/indonesia-urges-all-parties-to-exercise-restraint-ahead-of-south-china-sea-ruling. Accessed 22 Sep. 2016.

32. "Joint Verification Experiment." *Doomed to Cooperate*, https://lab2lab.stanford.edu/lab-lab/joint-verification-experiment. Accessed 15 Aug. 2021.

33. Ibid.

34. Ibid.

35. Ibid.

36. Song, Yann-huei. "A Marine Biodiversity Project in the South China Sea: Joint Efforts Made in the SCS Workshop Process." *The International Journal of Marine and Coastal Law*, vol. 26, no. 1, 2011, pp. 119–149. https://doi.org/10.1163/157180811X541413.

37. Borton, James. "Dispatches from the East Sea: Vietnam Showcases Island's Environmental Policy." *Geopolitical Monitor*, 29 June 2016, https://www.geopoliticalmonitor.com/dispatches-from-the-east-sea-vietnam-showcases-islands-environmental-policy/. Accessed 24 Oct. 2016.

38. At the airport in Manila before departing for an official visit to Japan, President Duterte stated that "I do not want to see any military man of any other nation [in the Philippines], except the Filipino soldier." He also added that the Philippines will survive without foreign investors who are squeamish about his bullish rhetoric.

39. McKirdy, Euan, and Kathy Quiano. "Philippines' Duterte to US: 'Do not make us your dogs.'" *CNN*, 25 October 2016, http://edition.cnn.com/2016/10/25/asia/duterte-us-comments/index.html. Accessed 7 Nov. 2016.

40. Parameswaran, Prashanth. "Malaysia Responds to China's South China Sea Intrusion." *The Diplomat*, 9 June 2015, http://thediplomat.com/2015/06/malaysia-responds-to-chinas-south-china-sea-intrusion/. Accessed 21 Sep. 2016.

41. Amindoni, Ayomi. "Indonesia Ready to Drive Peace, Prosperity in Asia: Jokowi." *The Jakarta Post*, 27 May 2016, http://www.thejakartapost.com/news/2016/05/27/indonesia-ready-to-drive-peace-prosperity-in-asia-jokowi.html. Accessed 21 Sep. 2016.

42. Nguyễn, Chu Hồi. "Phán Quyết Về Môi Trường Biển Đông Trong Vụ Philippin Kiện Trung Quốc." *Tạp chí Khoa học và Công nghệ Việt Nam*, vol. 11, 2016, pp. 61–64.

43. Walsh, Kathleen A. "The Blue Economy & Environmental Security: Implications for the South China Sea." Taiwan and the 2016 Elections: The Road Ahead, the Twenty-Fourth Annual Conference on Taiwan Affairs, Walker Institute of International and Area Studies University of South Carolina, US, 23–25 Sep. 2016.

44. Susskind, Lawrence. "Re: Follow-up from James Borton researcher/writer on environmental security in the South China Sea." Received by James Borton.

45. "Convention on the Law of the Sea." *UN General Assembly*, 10 Dec. 1982, http://www.refworld.org/docid/3dd8fd1b4.html. Accessed 22 Sep. 2016.

46. Mak, J. N. "Sovereignty in ASEAN and the Problem of Maritime Cooperation in the South China Sea." *Security and International Politics in the South China Sea*, edited by Sam Bateman, and Ralf Emmers, Routledge, 2009, p. 121.

47. "PNoy: JMSU with China, Vietnam 'Shouldn't Have Happened.'" *ABS-CBN News*, 4 Jan. 2011, http://news.abs-cbn.com/nation/01/04/11/pnoy-jmsu-china-vietnam-shouldnt-have-happened. Accessed 8 Nov. 2016.

48. Krasner, Stephen D. "Think again: Sovereignty." *Foreign Affairs*, 20 Nov. 2009, http://foreignpolicy.com/2009/11/20/think-again-sovereignty/. Accessed 8 Nov. 2016.

49. Xinhua. 2014. "Xi: There Is No Gene for Invasion in Our Blood." *China Daily*, 16 May 2014, https://www.chinadaily.com.cn/china/2014-05/16/content_17511204.htm.

50. Ataka, Hiroaki. "Geopolitics or Geobody Politics? Understanding the Rise of China and its Actions in the South China Sea." *Asian Journal of Peacebuilding*, vol. 4, no. 1, 2016, pp. 77–95. https://ipus.snu.ac.kr/eng/archives/ajp/volume-4-number-1-may-2016/research-article/geopolitics-or-geobody-politics-understanding-the-rise-of-china-and-its-actions-in-the-south-china-sea.

51. Glaser, Bonnie. "Re: Renewed Communication from James Borton, Environmental policy Writer working on SCS Science Diplomacy Policy Paper." Received by James Borton, 3 Nov. 2016.

52. The Devising Seminar on Arctic Fisheries summary report is available online at: dusp.mit.edu/sites/dusp.mit.edu/files/attachments/projects/AFDS_Summary Report.pdf. The Devising Seminar on Arctic Fisheries Stakeholder Assessment report is available online at: dusp.mit.edu/sites/dusp.mit.edu/files/attachments/ project/AFDS_StakeholderAssessment.pdf.

53. Susskind, Lawrence, and Danya Rumore. "Using Devising Seminars to Advance Collaborative Problem Solving in Complicated Public Policy Disputes." *Negotiation Journal*, vol. 31, no. 3, 2015, pp. 223–235.

54. Susskind, Lawrence, and Saleem H. Ali. *Environmental Diplomacy: Negotiating More Effective Global Agreements.* 2nd ed., New York, Oxford University Press, 2014.

References

"About." *China Dialogue Ocean*, chinadialogueocean.net/about/. Accessed 27 Jan. 2021.

"About Vietnam—Political System." *Socialist Republic of Vietnam Government Portal*, www.chinhphu.vn/portal/page/portal/English/TheSocialistRepublicOf Vietnam/AboutVietnam/AboutVietnamDetail?categoryId=10000103& articleId=10001578. Access 14 Aug. 2021.

"American Corners." *U.S. Embassy in the Democratic Republic of the Congo*, 13 Feb. 2019, cd.usembassy.gov/education-culture/american-spaces/american-corners/.

Bale, Rachel. "One of the World's Biggest Fisheries is On the Verge of Collapse." *National Geographic*, 29 Aug. 2016, https://news.nationalgeographic.com/2016/08/wildlife-south-china-sea-overfishing-threatens-collapse/.

Beech, Hannah. "China Limited the Mekong's Flow. Other Countries Suffered a Drought." *The New York Times*, 14 Apr. 2020, https://www.nytimes.com/2020/04/13/world/asia/china-mekong-drought.html.

Berkman, Paul Arthur, et al. "Science Diplomacy: A Crucible for Turning the Tide in the South China Sea," *East West Center*, 30 Sept. 2020, https://www.eastwestcenter.org/events/science-diplomacy-crucible-turning-the-tide-in-the-south-china-sea.

Borton, James. "Managing the South China Sea: Where Policy Meets Science." *The Diplomat*, 30 Apr. 2016, thediplomat.com/2016/04/managing-the-south-china-sea-where-policy-meets-science/.

Borton, James. "Marine Peace Park Plan Offers Promise for South China Sea." *Geopolitical Monitor*, 26 Oct. 2015, https://www.geopoliticalmonitor.com/marine-peace-park-plan-offers-promise-for-south-china-sea/.

Brownlie, Ian. *Principles of Public International Law*, 6th ed., Oxford, Oxford University Press, 2003.

Bruun, Ole. "Environmental Protection in the Hands of the State: Authoritarian Environmentalism and Popular Perceptions in Vietnam." *The Journal of Environment & Development*, vol. 29, no. 2, 2020, pp. 171–195. https://doi.org/10.1177%2F1070496520905625.

Buszynski, L. "ASEAN, grand strategy, and the South China Sea: between China and the United States." *Great Powers, Grand Strategies: The New game in the South China Sea*, edited by Anders Corr, Annapolis, Naval Institute Press, 2018, pp. 122–146.

Carson, Rachel. *Silent Spring*. Boston, Houghton Mifflin, 2002.

Carson, Rachel. *The Sea Around Us*. Oxford University Press, 1951.

Dang, Hung. "Pandend—Công đồng kết nối Y khoa phòng chống Covid-19." *Facebook*, 29 Jan. 2020, https://www.facebook.com/groups/155042985943381.

Dao, Toan. "Travel Restrictions a Blow for Philippines' Tuna Exports." *SeafoodSource Official Media*, 24 Mar. 2020, www.seafoodsource.com/news/supply-trade/travel-restrictions-a-blow-for-philippines-tuna-exports.

Dao, Toan. "Vietnam Sets Goal of USD 10.5 Billion in Seafood Exports for 2019." *SeafoodSource Official Media*, 15 Feb. 2019, www.seafoodsource.com/news/supply-trade/vietnam-sets-goal-of-usd-10-5-billion-in-seafood-exports-in-2019.

Dokken, Karin. "Environment, Security and Regionalism in the Asia-Pacific: Is Environmental Security a Useful Concept?" *The Pacific Review*, vol. 14, no. 4, Jan. 2001, pp. 509–30. https://doi.org/10.1080/09512740110087311.

Eyler, Brian, and Courtney Weatherby. "New Evidence: How China Turned Off the Tap on the Mekong River." *Stimson Center*, 13 Apr. 2020, https://www.stimson.org/2020/new-evidence-how-china-turned-off-the-mekong-tap/.

Fanell, J.E., "China's Maritime Sovereignty Campaign: Scarborough Shoal, the "New Spratly Islands", and Beyond." *Great Powers, Grand Strategies: The New Game in the South China Sea*, edited by Andres Corr, Naval Institute Press, 2018, pp. 106–121.

"Five Questions for: James Movick, Director-General, Pacific Islands Forum Fisheries Agency." *Business Advantage PNG*, 20 Feb. 2018, www.businessadvantagepng.com/five-questions-for-james-movick-director-general-pacific-islands-forum-fisheries-agency/.

Glaser, Bonnie S. "China's Island Building in the Spratly Islands: For What Purpose?" *Examining the South China Sea disputes: Papers from the Fifth Annual CSIS South China Sea Conference*, edited by Murray Hiebert, P. Nguyen, and Greg Poling, Washington, Center for Strategic & International Studies, 2015, pp. 31–41.

Gutierrez, Miren, et al. "China's Distant-Water Fishing Fleet: Scale, Impact and Governance." *ODI*, 1 June 2020, www.odi.org/publications/16958-china-s-distant-water-fishing-fleet-scale-impact-and-governance.

Hsu, Leo. "Praising James Agee." *Foto8*, 10 Dec. 2009, www.foto8.com/live/praising-james-agee/.

Huu, Khoa. "Mekong Delta Hit by Worst Drought Ever." *VNExpress*, 21 Mar. 2020, https://e.vnexpress.net/photo/news/mekong-delta-hit-by-worst-drought-ever-4071241.html#:~:text=In%202016%20it%20lost%208%2C000,12%20provinces%20and%20a%20city.

Huu, Thuan. "New Book Released to Pay Tribute to Truong Sa Martyrs." *Nhan Dan Online*, 6 July 2018, en.nhandan.org.vn/culture/item/6355202-new-book-released-to-pay-tribute-to-truong-sa-martyrs.html.

Impacts of Mekong Hydropower Dams and Climate Change in the Lower Mekong Delta. Mekong Environment Forum, April 27, 2020. Webinar.

"In Southern Vietnam, Hong Kong Paper Mill Renews Concern as Trial Run Nears." *Tuổi Trẻ News*, 24 Jun. 2016, https://tuoitrenews.vn/news/business/20160624/hong-kong-paper-mill-gives-rise-to-pollution-fears-in-southern-vietnam/22723.html.

Interview conducted with Professor Chu Manh Trinh in Cu Lao Cham June 2, 2016.

Islands and Rocks in the South China Sea: Post-Hague Ruling, edited by James Borton, Bloomington, Xlibris, 2017.

Johnson, Keith, and Dan De Luce. "Fishing Disputes Could Spark a South China Sea Crisis." *Foreign Policy*, 7 Apr. 2016, foreignpolicy.com/2016/04/07/fishing-disputes-could-spark-a-south-china-sea-crisis/.

Kimball, Jeffery. Review of *Road to Disaster: A New History of America's Descent into Vietnam*, by VanDeMark, Brian. *H-Net Reviews*, Jan. 2019, http://www.h-net.org/reviews/showrev.php?id=53340.

Kraft, Herman, et al. "Online Event: Taiwan and Indo-Pacific Regional Security Architecture Conference—Day 2." *Center for Strategic and International Studies*, 13 Jan. 2021, www.csis.org/events/online-event-taiwan-and-indo-pacific-regional-security-architecture-conference-day-2.

Law, Science & Ocean Management. Edited by Myron H. Nordqist, Ronan Long, Tomas Heidar, and John Norton Moore, Leiden/Boston: Martinus Nijhoff Publishers, 2007, pp. 271–293.

Le, Minh Khue. *The Distant Stars*, WordPress, 2016. *Zine Library*, https://zinelibrary.files.wordpress.com/2016/02/distantstars.pdf.

Le, Quang Binh, et al. *Reports on Movements to Protect 6700 Trees in Hanoi*. Hanoi, Hong Duc, 2015. http://isee.org.vn/Content/Home/Library/472/reports-on-movements-to-protect-6700-trees-in-hanoi..pdf.

Le, Quoc Vinh. "Chung Tay Chống Tin Giả—Fake News." *Facebook*, 4 Feb. 2020, https://www.facebook.com/groups/antifakenews.vn.

Le, Thu Mach. "The Emergence of Social Media and the Green Public Sphere in Vietnam." *Monash University*, Thesis, 2019. https://doi.org/10.26180/5dd268d62c266.

Leavenworth, Stuart. "In South China Sea Case, Ruling on Environment Hailed as Precedent." *The Christian Science Monitor*, 20 July 2016, www.csmonitor.com/World/Asia-Pacific/2016/0720/In-South-China-Sea-case-ruling-on-environment-hailed-as-precedent.

Li, Jianwei, et al. "Closing the Net Against IUU Fishing in the South China Sea: China's Practice and Way Forward." *Journal of International Wildlife Law & Policy*, vol. 18, no. 2, Jul. 2015, pp. 139–164. https://doi.org/10.1080/13880292.2015.1044799.

"Life Below Water." *The World Bank*, datatopics.worldbank.org/sdgatlas/archive/2017/SDG-14-life-below-water.html. Accessed 13 Aug. 2021.

Louv, Richard. *The Nature Principle: Human Restoration and the End of Nature-Deficit Disorder*. Chapel Hill, Algonquin Books, 2011.

Lovgren, Stefan. "Southeast Asia's Most Critical River is Entering Uncharted Waters." *National Geographic*, 31 Jan. 2020, https://www.nationalgeographic.com/science/article/southeast-asia-most-critical-river-enters-uncharted-waters#:~:text=-Originating%20in%20the%20icy%20headwaters,into%20the%20South%20China%20Sea.

Maefiti, John. *Pacific Islands Tuna Industry Association*, 10 July 2019, pacifictuna.org/.

"Major Paper Plant Halted due to Environmental Concerns." *VietnamNet*, 21 Nov. 2016, http://english.vietnamnet.vn/fms/environment/167335/major-paper-plant-halted-due-to-environmental-concerns.html.

McCalman, Iain. *The Reef: A Passionate History*. New York, Scientific American/Farrar, Straus and Giroux, 2014.

McManus, John, et al. "Science Diplomacy: A Crucible for Turning the Tide in the South China Sea." *East West Center*, 3 May 2016, https://www.eastwestcenter.org/events/environmental-security-crucible-in-the-south-china-sea.

"Medicine from the Sea." *Biological Magnetic Resonance Bank*, 7 Oct. 2020, bmrb.io/featuredSys/aspergillic_acid/.

Migration & International Legal Norms, edited by T. Alexander Aleinikoff, and Vincent Chetail, 2003.

"Mischief Reef." *Asia Maritime Transparency Initiative*, amti.csis.org/mischief-reef/.

Nash, Stephen E. "The Sound and Fury of the Huey Helicopter." *SAPIENS*, 21 Mar. 2017, www.sapiens.org/column/curiosities/the-sound-and-fury-of-the-huey-helicopter/.

"Nauru Agreement." *Pacific Islands Forum Fisheries Agency (FFA)*, 2020, www.ffa.int/nauru_agreement. Accessed 13 Aug. 2021.

Nguyen, Thao Thi Phuong, et al. "Fake News Affecting the Adherence of National Response Measures During the COVID-19 Lockdown Period: The Experience of Vietnam." *Frontiers in Public Health*, vol. 8, Sept. 2020, p. 589872. *PubMed Central*, https://doi.org/10.3389/fpubh.2020.589872.

Papanicolopulu, Irini. "The Duty to Rescue at Sea, in Peacetime and in War: A General Overview." *International Review of the Red Cross*, vol. 98, no. 2, 2016, pp. 491–514, https://library.icrc.org/library/docs/DOC/irrc-902-papanicolopulu.pdf.

Parker, Sam, and Gabrielle Chefitz. "Debt Book Diplomacy." *Belfer Center for Science and International Affairs*, 24 May 2018, www.belfercenter.org/publication/debtbook-diplomacy.

Parsons, Christopher, and Pierre-Louis Vézina. "Migrant Networks and Trade: The Vietnamese Boat People as a Natural Experiment," *IZA*, no. 10112, 2016, http://ftp.iza.org/dp10112.pdf.

Pham, Charlotte. "Boats in Vietnam." *Encyclopedia of the History of Science, Technology, and Medicine in Non-Western Cultures,* edited by Helaine Selin, Springer, Dordrecht, 2011.

Pham, Nguc Luy. *The Freedom Voyage of the Truong Xuan.* Translated by Phan Dam and Thien Vo-Dai, *Hearts of Freedom,* https://heartsoffreedom.org/wp-content/uploads/2019/09/The-Freedom-Voyage-of-the-Truong-Xuan.pdf.

Phan, Hai-Dang. "A Deluge of New Vietnamese Poetry." *Asymptote*, www.asymptote-journal.com/special-feature/haidang-phan-on-contemporary-vietnamese-poets/. Accessed 14 Aug. 2021.

Pitcher, Tony, et al. "Not Honouring the Code." *Nature*, vol. 457, no. 7230, Feb. 2009, pp. 658–59, https://doi.org/10.1038/457658a.

Quang, Nguyen Minh. "Is Vietnam in for Another Devastating Drought?" *The Diplomat*, 8 Feb. 2017, https://thediplomat.com/2017/02/is-vietnam-in-for-another-devastating-drought/.

Quang, Nguyen Minh, and Dao Ngoc Canh. "Saving Vietnam's Floating Markets." *The Diplomat*, 21 Mar. 2017, https://thediplomat.com/2017/03/saving-vietnams-floating-markets/.

Raven, Peter H., and Lynn Margulis. "The Herbal of Rumphius." *American Scientist*, 6 Mar. 2018, www.americanscientist.org/article/the-herbal-of-rumphius.

Reuters. "Payments Due: Pacific Islands in the Red as Debts to People's Republic of China Mount." *Indo-Pacific Defense Forum*, 27 Aug. 2018, ipdefenseforum.com/2018/08/payments-due-pacific-islands-in-the-red-as-debts-to-peoples-republic-of-china-mount/.

Roberts, Tyson R. "Just Another Dammed River? Negative Impacts of Pak Mun Dam on Fishes of the Mekong Basin." *Natural History Bulletin of the Siam Society*, vol. 41, 1993, pp. 105–133.

Roy, Nalanda. *Navigating Uncertainty in the South China Sea Disputes: Interdisciplinary Perspectives.* Singapore, World Scientific, 2020.

"Sea of Troubles." *The Economist,* 2 May 2015, www.economist.com/leaders/2015/05/02/sea-of-troubles.

Seidel, H., and Padma N. Lal. *Economic Value of the Pacific Ocean to the Pacific Island Countries and Territories.* Gland, Switzerland, IUCN, 2010. https://www.iucn.org/sites/dev/files/import/downloads/economic_value_of_the_pacific_ocean_to_the_pacific_island_countries_and_territories_p.pdf.

Sheko, Tania. "Warsan Shire: 'Home.'" *Poem of the Day,* 12 Sept. 2017, http://medium.com/poem-of-the-day/warsan-shire-home-46630fcc90ab.

Soong, Keryea. Oceanography Department National Sun Yat-Sen University, 17 Feb. 2021.

"South China Sea." *International—U.S. Energy Information Administration (EIA),* 7 Feb. 2013, www.eia.gov/international/analysis/regions-of-interest/South_China_Sea.

Stobutzki, I. C., et al. "Decline of Demersal Coastal Fisheries Resources in Three Developing Asian Countries." *Fisheries Research,* vol. 78, no. 2, May 2006, pp. 130–42. *ScienceDirect,* https://doi.org/10.1016/j.fishres.2006.02.004.

Stout, David. "South China Sea: The Last Time China and Vietnam Fought, It Was Hell." *Time,* 15 May 2014, time.com/100417/china-vietnam-sino-vietnamese-war-south-china-sea/.

Teh, Louise S. L., et al. "What Is at Stake? Status and Threats to South China Sea Marine Fisheries." *Ambio,* vol. 46, no. 1, Feb. 2017, pp. 57–72. *Springer Link,* https://doi.org/10.1007/s13280-016-0819-0.

"Temple of Literature, Hanoi." *Wikipedia,* 6 Feb. 2021, en.wikipedia.org/wiki/Temple_of_Literature,_Hanoi.

"The Link Between Effective Fisheries Management and Ending Harmful Subsidies." *The Pew Charitable Trusts,* 5 Apr. 2019, www.pewtrusts.org/en/research-and-analysis/issue-briefs/2019/04/the-link-between-effective-fisheries-management-and-ending-harmful-subsidies.

The South China Sea: Challenges and Promises, edited by James Borton, Bloomington, Xlibris, 2015.

Thi, Hue Hoang, and HaHoang Thi Hong. "Acculturation in Vietnamese Contemporary Literature." *International Journal of Communication and Media Studies (IJCMS),* vol. 6, no. 3, June 2016, https://papers.ssrn.com/sol3/papers.cfm?abstract_id=2835456.

"United Nations Convention on the Law of the Sea." *The United Nations,* 10 Dec. 1982, https://www.un.org/depts/los/convention_agreements/texts/unclos/unclos_e.pdf. Accessed 15 Aug. 2021.

University of British Columbia. "Chinese Foreign Fisheries Catch 12 Times More than Reported, Study Shows." *ScienceDaily,* 3 Apr. 2013, www.sciencedaily.com/releases/2013/04/130403104210.htm.

US Department of Commerce. *National Oceanic and Atmospheric Administration*, 2021, www.noaa.gov/.

Varley, Kevin, et al. "Fight over Fish Fans a New Stage of Conflict in South China Sea." *Bloomberg*, 1 Sept. 2020, www.bloomberg.com/graphics/2020-dangerous-conditions-in-depleted-south-china-sea/.

"Vietnam, World Bank Sign $560 million to Support Mekong Delta Urban Development and Climate Resilience." *The World Bank,* 11 July 2016, https://www.worldbank.org/en/news/press-release/2016/07/11/vietnam-world-bank-sign-560-million-to-support-mekong-delta-urban-development-and-climate-resilience.

Vu, N.A. "Grassroots Environmental Activism in an Authoritarian Context: The Trees Movement in Vietnam." *Voluntas: International Journal of Voluntary and Nonprofit Organizations*, vol. 28, 2017, pp. 1180–1208. https://doi.org/10.1007/s11266-017-9829-1.

Wang, Bin. "The Outlook for the Establishment and Management of Marine Protected Area Network in China." *International Journal of Geoheritage and Parks*, vol. 6, no. 1, June 2018, pp. 32–42. *ScienceDirect*, https://doi.org/10.17149/ijg.j.issn.2210.3382.2018.01.003.

Wenzel, Lauren et al. "Marine Reserves in the United States." *National Marine Protected Areas Center*, 2014. http://marineprotectedareas.noaa.gov/pdf/helpful-resources/factsheets/reserves-factsheet2014.pdf.

"Western & Central Pacific Fisheries Commission (WCPFC)." *Western & Central Pacific Fisheries Commission*, 19 June 2004, www.wcpfc.int/.

Winston, Rachel A., and Ishika Sachdeva. *Raging Waters in the South China Sea: What the Battle for Supremacy Means for Southeast Asia*. Irvine, Lizard Publishing, 2020.

Zhang, Jian. "China's Role in the Pacific Islands Region," *APCSS*, 2015. https://apcss.org/wp-content/uploads/2015/08/C3-China-Pacific-Zhang.pdf.

Index

CPSIA information can be obtained
at www.ICGtesting.com
Printed in the USA
BVHW041830080322
630930BV00017B/169